the
UNIVERSITY
of
GREENWICH

ETHYLENE

ADVANCES IN AGRICULTURAL BIOTECHNOLOGY

Already published in this series

Akazawa T., et al., eds: The New Frontiers in Plant Biochemistry. 1983.
ISBN 90-247-2829-0

Gottschalk W. and Müller H.P., eds: Seed Proteins: Biochemistry, Genetics, Nutritive Value. 1983. ISBN 90-247-2789-8

Marcelle R., Clijsters H. and Van Poucke M., eds: Effects of Stress on Photosynthesis. 1983. ISBN 90-247-2799-5

Veeger C. and Newton W.E., eds: Advances in Nitrogen Fixation Research. 1984. ISBN 90-247-2906-8

Chinoy N.J., ed: The Role of Ascorbic Acid in Growth, Differentiation and Metabolism of Plants. 1984. ISBN 90-247-2908-4

Witcombe J.R. and Erskine W., eds: Genetic Resources and Their Exploitation – Chickpeas, Faba beans and Lentils. 1984. ISBN 90-247-2939-4

Sybesma C., ed: Advances in Photosynthesis Research. Vols. I-IV. 1984.
ISBN 90-247-2946-7

Sironval C., and Brouers M., eds: Protochlorophyllide Reduction and Greening. 1984. ISBN 90-247-2954-8

Ethylene

Biochemical, Physiological and Applied Aspects

An International Symposium, Oiryat Anavim, Israel
held January 9 12 1984

edited by

YORAM FUCHS
EDO CHALUTZ

Department of Fruit and Vegetable Storage
Agricultural Research Organization
The Volcani Center
Bet Dagan, Israel

1984 **MARTINUS NIJHOFF/DR W. JUNK PUBLISHERS**
a member of the KLUWER ACADEMIC PUBLISHERS GROUP
THE HAGUE / BOSTON / LANCASTER

Distributors

for the United States and Canada: Kluwer Academic Publishers, 190 Old Derby Street, Hingham, MA 02043, USA
for the UK and Ireland: Kluwer Academic Publishers, MTP Press Limited, Falcon House, Queen Square, Lancaster LA1 1RN, England
for all other countries: Kluwer Academic Publishers Group, Distribution Center, P.O. Box 322, 3300 AH Dordrecht, The Netherlands

Library of Congress Cataloging in Publication Data

```
Main entry under title:

Ethylene: biochemical, physiological, and applied
   aspects.

   (Advances in agricultural biotechnology)
   "An international symposium, Qiryat Anavim, Israel,
January 9-12, 1984, sponsored by Israel Scientific
research conferences, national council for research and
development"--
   Includes index.
   1. Plants, effect of ethylene on--Congresses.
2. Ethylene--Synthesis--Congresses. 3. Plant hormones
--Congresses. I. Fuchs, yoram. II Chaluts, edo.
III. Israel scientific research conferences.
IV. Series.
QK898.E8E86 1984      582'.01927        84-10175
ISBN 90-247-2984-X
```

ISBN 90-247-2984-x (this volume)
ISBN 90-247-2790-1 (series)

Book information

Sponsored by Israel Scientific Research Conferences, National Council for Research and Development
Also supported by U.S. – Israel Binational Agricultural Research and Development Fund (BARD)

Copyright

PRINTED IN THE NETHERLANDS

This symposium is dedicated to the memory of our
colleague and friend - Morris Lieberman

Morris Lieberman died on January 18, 1982. At the time of his death he was an active and respected scientist, in charge of the Plant Hormone Laboratory of the Agricultural Research Service, the United States Department of Agriculture, at Beltsville, Maryland. Among the many contributions he made to the science of plant physiology, his most significant work was in the field of ethylene, which began with the discovery, together with Dr. Mapson and associates, of the amino acid methionine as the precursor of ethylene in higher plants. His later work with Owens, on the discovery of the specific inhibition of ethylene biosynthesis by rhizobitoxine and its analog aminoethoxyvinylglycine (AVG), opened new avenues in ethylene research. Morris Lieberman, with others in his laboratory, pioneered studies of the interaction of ethylene with other plant hormones, of the association of ethylene biosynthesis with membranes, and of the role of ethylene in leaf senescence.

Dr. Lieberman's personal integrity, warm personality and scientific reputation attracted scientists to his laboratory from all over the world. Every one enjoyed his informal, kind disposition and friendly hospitality. Morris had a great enthusiasm for science, coupled with an objective judgement. Combining a fine sense of humor and a charming personality, he was always available for professional discussions, with an open and helpful attitude.

The idea of organizing an international symposium on ethylene in Israel was born 7 years ago at a dinner party which some of us attended with Morris Lieberman, during his last visit to Israel. Ethylene has been of interest to more than a few Israeli scientists, many of whom had a close relationship with Morris or had actually studied or worked in his laboratory. Following his untimely death we felt that organizing a symposium on ethylene dedicated to his memory would indeed be the most appropriate way of honoring Morris Lieberman, a great scientist and a dear friend.

Yoram Fuchs and Edo Chalutz

P R E F A C E

The plant hormone ethylene plays an important regulatory role in the physiology of plants and, in particular, the senescence and postharvest physiology of fruits and vegetables. During the last two decades many exciting developments have taken place in the research on ethylene biosynthesis, its mode of action and its physiological role in plants, but no scientific meeting devoted entirely to ethylene had been organized. Therefore, we thought that a symposium in this field would offer an opportunity to bring together many scientists interested in ethylene, to report on new findings and developments, to exchange ideas and to discuss future directions of ethylene research. Indeed, this opportunity was fully exploited at this recent meeting.

We wish to thank the Israel National Council for Research and Development (NCRD) for recognizing the need and importance of sponsoring the symposium, for inviting some of the participants, and for providing the framework, organization and facilities. Special thanks are due to Mrs. Shulamit Cahana and Mr. David Birenbach, who attended to all the organization details, and to Dr. Baruch Eyal, Head of the International Affairs Division and Dr. Miriam Waldman, Coordinator of Biological Research of the NCRD, for helping in planning and organizing the meeting. We also thank Mr. Yair Guron, Director of the U.S. - Israel Binational Agricultural Research and Development Fund (BARD) for his encouragement and advice, and the BARD authorities for their generous financial support.

We express our appreciation to our colleagues on the Organizing Committee for their collaboration and all the participants in the Symposium who helped make it an interesting, fruitful and pleasant event.

Yoram Fuchs and Edo Chalutz

Bet Dagan, Israel

April 1984

ORGANIZING COMMITTEE

Fuchs, Y.	ARO, The Volcani Center, Bet Dagan
Atzmon, D.	Weizmann Institute of Science, Rehovot
Cahana, S.	National Council for Research & Development, Jerusalem
Chalutz, E.	ARO, The Volcani Center, Bet Dagan
Eyal, B.	National Council for Research & Development, Jerusalem
Goren, R.	Faculty of Agriculture, Hebrew University of Jerusalem, Rehovot
Halevy, A.H.	Faculty of Agriculture, Hebrew University of Jerusalem, Rehovot
Leshem, Y.	Bar-Ilan University, Ramat Gan
Mayak, S.	Faculty of Agriculture, Hebrew University of Jerusalem, Rehovot
Mizrahi, Y.	Ben-Gurion University of the Negev, Be'er Sheva
Waldman, M.	National Council for Research & Development, Jerusalem

C O N T E N T S

THE FORMATION OF ETHYLENE FROM 1-AMINOCYCLOPROPANE-1-CARBOXYLIC ACID

S.F. YANG

Vegetable Crops Dept., University of California, Davis,
CA 95616, USA

Since Adams and Yang (1) demonstrated that ethylene is biosynthesized in apple tissue via the following sequence: methionine --> SAM --> ACC --> ethylene, this pathway has been shown to operate throughout the diversity of higher plant tissues. Although EFE, which catalyzes the terminal step in the sequence, has not been well characterized _in vitro_, it is clear from the _in vivo_ data that the reaction is oxygen-dependent, inhibited by various agents, activated by CO_2, and exhibits stereoselectivity toward ACC analogs. In this paper I shall summarize the present status of knowledge pertaining to the mechanism and regulation of this reaction in various plant systems.

IN VIVO CHARACTERISTICS OF EFE

When ACC was applied to a number of plant tissues, a marked increase in ethylene production was observed (2,3). This suggests that EFE is constitutive in these tissues. In others, such as preclimacteric fruit tissue (1) or young petals of carnation flower (4), EFE activity is low, but increases markedly as they undergo ripening or the senescence process. EFE can also be induced by certain external factors, such as wilting stress and excision.

It has long been recognized that various lipophilic compounds, such as phosphatidylcholine, Tween 20 and Triton X-100, and osmotic shock treatment, all of which could modify membrane structure, greatly reduce the rate of ethylene synthesis in plant tissues. Moreover, when those

----

Abbreviations: ACC=1-aminocyclopropane-1-carboxylic acid; AIB= α-aminoisobutyric acid; AEC=1-amino-2-ethylcyclopropane-1-carboxylic acid; DCMU=3-(3,4-dichlorophenyl)-1,1-dimethylurea; DNP=2,4-dinitrophenol; EFE=ethylene forming enzyme which converts ACC to ethylene; SAM=S-adenosylmethionine.

Y. Fuchs and E. Chalutz (eds.) Ethylene: Biochemical, Physiological and Applied Aspects.
ISBN 90-247-2984-X. Printed in The Netherlands
©1984, Martinus Nijhoff/Dr W. Junk Publishers, The Hague.

tissues, which are producing ethylene actively, are homogenized, the
ethylene-forming capability is totally lost. These observations lead to
the suggestion that the ethylene-forming system is highly structured and
requires membrane integrity (5). Later work showed that the step at which
lipophilic compounds and osmotic shock inhibited ethylene production was
the conversion of ACC to ethylene. The uncouplers of oxidative
phosphorylation, such as DNP, are potent inhibitors of ethylene production
(5). Low concentrations of DNP inhibited ethylene synthesis by blocking
the conversion of ACC to ethylene without affecting the conversion of
methionine to SAM (6), a step which requires ATP. The mechanism by which
uncouplers inhibit EFE is, however, unclear. One explanation is that ATP
is required for the conversion of ACC to ethylene. Alternatively, DNP may
disrupt membrane structure, the integrity of which may be essential for
EFE. Recently, John (7) has suggested that the generation of ethylene
from ACC is coupled to a transmembrane flow of protons from the outside to
the inside of the plasma membrane. This model explains the strict
dependence of ethylene biosynthesis on membrane integrity and the marked
inhibition by the protonophore, DNP. Although this is an interesting
hypothesis, experimental evidence is lacking.

Free radical inhibitors, such as n-propyl gallate, inhibit ethylene
production in various plant tissues by inhibiting the conversion of ACC to
ethylene (8). These observations lead to the contention that the
conversion of ACC to ethylene is mediated by a free radical reaction.
However, the concentration of these inhibitors required to inhibit
ethylene production is rather high. Other compounds or factors which
effectively inhibit the conversion of ACC to ethylene are Co^{2+} (9),
temperature above 40 C (6) and polyamines (10). Among the structural
analogs of ACC tested, only AIB significantly and competitively inhibited
ethylene production, although the inhibition was not as marked as other
inhibitors previously discussed (11). In contrast to the general view
that AIB is a metabolically inert amino acid, [carboxyl-^{14}C]AIB has
been recently shown to be metabolized to $^{14}CO_2$ and N-malonyl-
[^{14}C]AIB in mungbean hypocotyls (12). Both the decarboxylation of

AIB and the conversion of ACC to ethylene require oxygen and are inhibited
by the same inhibitors, suggesting that oxidative decarboxylation of AIB
and the oxidative degradation of ACC are catalyzed by the same enzyme,
EFE. Thus, AIB competes with ACC for the site of EFE, and is itself
degraded to CO_2 and other products.

REGULATION BY LIGHT AND CARBON DIOXIDE

Recently many investigators have observed that light markedly
inhibited ethylene production by various green leaf tissues in enclosed
containers. Since this inhibitory effect of light was not observed in
etiolated tissues, was not reversible by red, far-red light treatments,
and was inhibited by DCMU, an inhibitor of photosynthetic electron
transport, it was concluded that the light inhibition was related to the
photosynthetic system. Gepstein and Thimann (13) were the first to report
that the conversion of ACC to ethylene was inhibited by light. Since
CO_2 is known to promote ethylene production in leaf tissues (14),
Grodzinski et al (15) and Kao & Yang (16) reasoned that the inhibition of

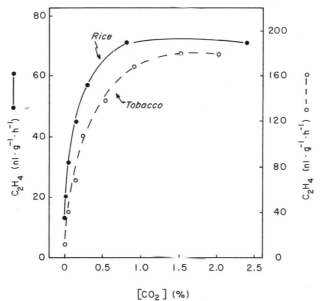

FIGURE 1. Effect of CO_2 concentration on ACC-dependent ethylene
production in rice leaf segments and tobacco leaf discs under light. From
Kao and Yang (16).

ethylene evolution by light might result from a decrease in internal CO_2 concentration. In excised rice leaves light substantially inhibited the endogenous ethylene production, but when CO_2 was added into the incubation flask, the rate of ethylene production in the light increased markedly, to a level which was even higher than that produced in the dark (16). Carbon dioxide, however, had no appreciable effect on leaf segments incubated in the dark. Similar results were obtained with ACC-treated tobacco leaf discs. The concentrations of CO_2 giving half-maximal activity was 0.06% and 0.18% for rice and tobacco leaves, respectively (Fig. 1). Thus, it is the CO_2 metabolism rather than light per se, which regulates the conversion of ACC to ethylene. The modulation of ACC conversion to ethylene by CO_2 or light was rapid and reversible, indicating that CO_2 regulates the activity, but not the synthesis of EFE (16). The mechanism by which CO_2 modulates the conversion of ACC to ethylene is not understood.

REGULATION BY ETHYLENE

Autocatalysis of ethylene production is a characteristic feature of ripening fruits and other senescing tissues in which a massive increase in ethylene production can be triggered by exposure to ethylene. Since preclimacteric (unripe) fruits lack both ACC synthase and EFE, a massive increase in ethylene production requires development of both enzymes. However, when green tomato fruits were treated with ethylene for a short period (18 h), there was no increase in ACC content or in ethylene production rate but the tissue's ability to convert ACC to ethylene (EFE) increased markedly (Table 1). These data indicate that when preclim-acteric fruit tissues are exposed to ethylene the increase in EFE precedes the increase in ACC synthase. Whether or not this is also true during the natural ripening of fruits remains to be clarified. Ethylene has also been shown to stimulate the conversion of ACC to ethylene in excised cantaloupe fruit and citrus leaf.

REGIOSPECIFICITY

Although the ACC molecule possesses two enantiotopic methylene groups, they are not geometrically equivalent and can be distinguished by

Table 1. Promotion of EFE development by ethylene in preclimacteric
tomato fruit.

   Fruits of immature green tomato were treated with air or 10 μl
l^{-1} ethylene for 18 h. Discs (0.5 cm diameter) were then prepared
from the pericarp tissue, and ethylene produced during and ACC content
at the end of a 6-h incubation period were determined. The ethylene
produced in the presence of 2 mM ACC during the 6-h incubation period
was taken as an index of EFE (Y. Liu, unpublished results)

Pretreatment	ACC	C_2H_4	EFE
	nmol g^{-1}	nmol g^{-1}	nmol g^{-1}
Air	1.65	1.49	8.77
Ethylene	0.41	1.34	48.44

a regiospecific enzyme. Ethyl substitution of each of the 4 methylene
hydrogens results in 4 stereoisomers of AEC. If ACC conversion to
ethylene by plant tissues proceeds in regiospecific fashion, Hoffman et al
(17) have reasoned that these four stereoisomers of AEC may not be converted into 1-butene with equal efficiency. In apple and etiolated
mungbean hypocotyls, (1R, 2S)-AEC was preferentially converted to
1-butene. By chemical oxidation using NaOCl, in contrast, all AEC isomers
were converted with nearly equal efficiency to butene. ACC and AEC appear
to be degraded by the same enzyme since both reactions are inhibited to
the same extent by nitrogen atmosphere or by Co^{2+}, and since, when
both substrates are present simultaneously, each acts as an inhibitor with
respect to the other. These observations indicate that the enzyme
converting ACC to ethylene exhibits regiospecificity.

   Soon after ACC was established as the immediate precursor of
ethylene, Konze and Kende (18) reported an enzyme extract from etiolated
pea seedlings capable of converting ACC to ethylene. Many similar systems
have since been reported, including carnation microsomes (4), pea microsomes (19), pea mitochondria (20), IAA-oxidase and peroxidase (21).
Although these systems are oxygen-dependent, heat-denaturable and inhibited by the radical scavengers, there are many characteristics which do

not resemble those of the natural _in vivo_ system. McKeon and Yang (22)
have compared the regiospecificity, with regard to the conversion of AEC
isomers to 1-butene, by pea epicotyls and by the pea epicotyl extract.
While the pea epicotyls displayed regiospecificity as observed in mungbean
hypocotyls and apple tissue, and exhibited high affinity for ACC with a
Km of about 66 μM with respect to the internal ACC concentration, the pea
homogenate did not differentiate between AEC isomers (Table II) and
exhibited very low affinity for ACC (reported Km for the pea enzyme ranged
from 15 to 400 mM).

Table II. Conversion of AEC isomers (3 mM) to 1-butene by pea stem
segments or pea stem homogenate. From McKeon and Yang (22)

AEC isomer	1-Butene production (nmol $g^{-1} h^{-1}$)	
	Pea Stem	Pea Stem Homogenate
(1R,2S)-AEC	1.26	0.56
(1S,2R)-AEC	0.03	0.57
(1S,2S)-AEC	0.01	0.20
(1R,2R)-AEC	0.02	0.24

Moreover, the pea enzyme required Mn^{2+}, and was very sensitive to
inhibition by EDTA and mercaptoethanol, whereas the _in vivo_ system was
not. These data do not support the view that the reported enzymic systems
are responsible for the conversion of ACC to ethylene _in vivo_. It should
be noted that ACC can be converted to ethylene chemically by various
oxidants including oxidative free radicals (23). A simple explanation
accounting for the low affinity and lack of regiospecificity in these
enzyme systems is that the physiological enzyme (EFE) catalyzes directly
the oxidation of ACC yielding ethylene, whereas the reported enzymes
catalyze the activation of molecular oxygen probably to free radicals,
such as O_2^- and OH, which in turn react with ACC nonenzymatically
to form ethylene. Such reaction mechanisms are in agreement with the
observation that the _in vitro_ systems have high Km values but are
saturable with ACC, because as the concentrations of ACC are increased,
the rate-limiting reaction shifts to the reaction of oxygen activation,
the rate of which is independent of ACC concentration. Recently Guy and
Kende (24) observed that the vacuole fraction isolated from pea
protoplasts accounted for 80% of the protoplast ethylene production.

Unlike the pea enzyme system, the vacuoles resemble the _in vivo_ system in that they differentiate between AEC isomers, have an apparent Km of 70 μM for ACC, are sensitive to Co^{2+} inhibition, and are very sensitive to membrane disruption. Thus, intact vacuoles appear to be the smallest biological units which possess properties characteristic of EFE.

MECHANISM OF THE REACTION

Although the enzyme system responsible for the conversion of ACC to ethylene remains to be characterized, some progress has been made with respect to the reaction mechanism. Analogous to the chemical oxidation, we have previously proposed that ACC can be oxidized by a hydroxylase to N-hydroxy-ACC, which is a nitrenium equivalent, or by a dehydrogenase to form the nitrenium intermediate, which is then fragmented into ethylene and cyanoformic acid (25). Cyanoformic acid, is very labile and degraded spontaneously to CO_2 (derived from the carboxyl group of ACC) and HCN (derived from C-1 of ACC). The overall reaction represents a two-electron oxidation.

The validity of this pathway is supported by the recent observation of Peiser et al (26). They showed that [1-^{14}C]ACC was primarily converted into [4-^{14}C]asparagine, a hydrated product of β-[^{14}C]cyanoalanine, in mungbean hypocotyls, and into β-[^{14}C]cyanoalanine and γ-glutamyl-α-[^{14}C]cyanoalanine in _Vicia sativa_ epicotyls, in amounts similar to the ethylene produced. When Na^{14}CN was administered to mungbean hypocotyls or _Vicia sativa_ epicotyls, it was similarly incorporated into asparagine in mungbean, and into β-cyanoalanine and its conjugate in _Vicia sativa_. When [carboxyl-^{14}C]ACC was administered into those plant tissues, CO_2 was produced in an amount equivalent to the amount of ethylene produced. Thus, in the conversion of ACC to ethylene, the carboxyl group yields

CO_2, and C-1 gives off HCN. Since HCN is toxic to plants, it is
pertinent to ask whether plants have ample capacity to detoxify the HCN
thus formed. β-Cyanoalanine synthase, which catalyzes the formation of
β-cyanoalanine from cysteine and HCN, is widely distributed in higher
plants. While ethylene production rate in higher plants ranges from 0 to
0.2 nmol g^{-1} min^{-1}, β-cyanoalanine synthase activity ranges from 4
to 1000 nmol g^{-1} min^{-1} (27). Recently, Adlington et al (28) have
fed [cis- 2,3-$^2H_2$]ACC to apple slices and observed that it
yielded a mixture of equal amounts of cis- and trans-[1,2-$^2H_2$]
ethylene. This indicates that the configuration of hydrogens is lost
during the conversion of ACC to ethylene by apple tissue. In contrast to
the biosynthetic results, the chemical oxidation of ACC to ethylene with
NaOCl results in complete retention of ethylene configuration. These data
indicate that NaOCl oxidation of ACC to ethylene may proceed by a
concerted elimination mechanism, but the biosynthesis proceeds instead by
a stepwise mechanism involving an intermediate that allows scrambling of
the hydrogens in the ring. Most recently, Pirrung (29) has studied the
electrochemical oxidation of [cis-2,3-$^2H_2$]ACC, which results in
ethylene formation with loss of stereochemistry, as observed in plant
tissue. He has therefore suggested that in vivo oxidation of ACC
proceeds in two sequential one-electron oxidation reactions, and the
initial oxidation yields amine radical intermediates which undergo rapid
ring-opening and loss of stereochemistry. This route is analogous to
proposed amine radical intermediates in amine oxidase catalysis. These
observation suggest that EFE is probably not a hydroxylase but a
flavo-oxidase (Enz-F), which utilize molecular oxygen as its ultimate
electron acceptor.

ACKNOWLEDGEMENTS

Our work cited herein has been supported by research grants from NSF (PCM-8114933) and BARD (I-145-79 and I-221-80).

REFERENCES

1. Adams DO, Yang SF. 1979. Ethylene biosynthesis: identification of 1-aminocyclopropane-1-carboxylic acid as an intermediate in the conversion of methionine to ethylene. Proc Natl Acad Sci 76:170-174.
2. Cameron AC, Fenton CAL, Yu YB, Adams DO, Yang SF. 1979. Increased production of ethylene by plant tissues treated with 1-amino-cyclopropane-1-carboxylic acid. Hortscience 14:178-180.
3. Lurssen K, Naumann K, Schroder R. 1979. 1-Aminocyclopropane-1-carboxylic acid- an intermediate of the ethylene biosynthesis in higher plants. Z. Pflanzenphysiol 92:285-294.
4. Mayak S, Legge RL, Thompson JE. 1981. Ethylene formation from 1-aminocyclopropane-1-carboxylic acid by microsomal membranes from senescing carnation flowers. Planta 153:49-55.
5. Lieberman M. 1979. Biosynthesis and action of ethylene. Ann Rev Plant Physiol 30:533-591.
6. Yu YB, Adams DO, Yang SF. 1980. Inhibition of ethylene production by 2,4-dinitrophenol and high temperature. Plant Physiol 66:286-290.
7. John P. 1983. The coupling of ethylene biosynthesis to a transmembrane electrogenic proton flux. FEBS Lett 152:141-143.
8. Apelbaum A, Wang SY, Burgoon AC, Baker JE, Lieberman M. 1981. Inhibition of the conversion of 1-aminocyclopropane-1-carboxylic acid to ethylene by structural analogs, inhibitors of electron transfer, uncouplers of oxidative phosphorylation, and free radical scavengers. Plant Physiol 67:74-79.
9. Yu YB, Yang SF. 1979 Auxin-induced ethylene production and its inhibition by aminoethoxyvinylglycine and cobalt ion. Plant Physiol 64:1074-1077.
10. Suttle JC. 1981. Effect of polyamines on ethylene production. Phytochemistry 20:1477-1480.
11. Satoh S, Esashi Y. 1983. Alpha-aminoisobutyric acid, propyl gallate and cobalt ion and the mode of inhibition of ethylene production by cotyledonary segments of cocklebur seeds.
12. Liu Y, Su L, Yang SF. 1984. Metabolism of α-aminoisobutyric acid in mungbean hypocotyls in relation to metabolism of 1-aminocyclopropane -1-carboxylic acid (submitted).
13. Gepstein S, Thimann KV. 1980. The effect of light on the production of ethylene from 1-aminocyclopropane-1-carboxylic acid by leaves. Planta 149:196-199.
14. Bassi PK, Spencer MS. 1982. Effect of carbon dioxide and light on ethylene production in intact sunflower plants. Plant Physiol 69:1222-25.
15. Grodzinski B, Boesel I, Horton RF. 1982. Ethylene release from leaves of Xanthium strumarium L. and Zea mays L. J Exp Bot 33:344-354.

16. Kao CH, Yang, SF. 1982. Light inhibition of the conversion of 1-aminocyclopropane-1-carboxylic acid to ethylene in leaves is mediated through carbon dioxide. Planta 155:261-266.

17. Hoffman NE, Yang SF, Ichihara A, Sakamura S. 1982. Stereospecific conversion of 1-aminocyclopropanecarboxylic acid to ethylene by plant tissues. Conversion of stereoisomers of 1-amino-2-ethylcyclopropanecarboxylic acid to 1-butene. Plant Physiol 70:195-199.

18. Konze JR, Kende H. 1979. Ethylene formation from 1-aminocyclopropane-1-carboxylic acid in homogenates of etiolated pea seedlings. Planta 146:293-301.

19. McRae DG, Baker JE, Thompson JE. 1982. Evidence for involvement of the superoxide radical in the conversion of 1-aminocyclopropane-1-carboxylic acid to ethylene by pea microsomal membranes. Plant Cell Physiol 23:375-383.

20. Vinkler C, Apelbaum A. 1983. Ethylene formation from 1-aminocyclopropane-1-carboxylic acid in plant mitochondria. FEBS Letters 162:252-256.

21. Vioque A, Albi MA, Vioque B. 1981. Role of IAA-oxidase in the formation of ethylene from 1-aminocyclopropane- 1-carboxylic acid. Phytochemistry 20:1473-1475.

22. McKeon TA, Yang SF. 1983. A comparison of the conversion of 1-amino-2-ethylcyclopropane-1-carboxylic acid stereoisomers to 1-butene by pea epicotyls and by cell free system. Planta (in press).

23. Legge RL, Thompson JE, Baker JE. 1982. Free radical-mediated formation of ethylene from 1-aminocyclopropane-1-carboxylic acid: a spin-trap study. Plant Cell Physiol 23:171-177.

24. Guy M, Kende H. 1984. Conversion of 1-aminocyclopropane-1-carboxylic acid to ethylene by isolated vacuoles of _Pisum sativum_ L. Planta (in press).

25. Yang SF. 1981. Biosynthesis of ethylene and its regulation. In: Recent Advances in the biochemistry of fruit and vegetables, eds., J Friend, MJC Rhodes, pp. 89-106, Academic Press, London.

26. Peiser G, Wang TT, Hoffman N, Yang SF. 1983. Evidence for $^{14}CN^-$ formation from $[1-^{14}C]ACC$ during _in vivo_ conversion of ACC to ethylene. Plant Physiol Suppl 72:No. 203 (Abstr).

27. Conn EE, Miller JM. 1980. Metabolism of hydrogen cyanide by higher plants. Plant Physiol 65:1199-1202.

28. Adlington RM, Baldwin JE, Rawlings BJ. 1983. On the stereochemistry of ethylene biosynthesis. J Chem Soc Chem Commun 1983:290-292.

29. Pirrung, MC. 1983. Ethylene biosynthesis II. Stereochemistry of ripening, stress, and model reactions. J Am Chem Soc 105:7207.

THE BIOCHEMISTRY AND PHYSIOLOGY OF 1-AMINOCYCLOPROPANE-1-CARBOXYLIC ACID CONJUGATION

N. AMRHEIN, U. DORZOK, C. KIONKA, U. KONDZIOLKA, H. SKORUPKA and S. TOPHOF

Ruhr-Universität Bochum, 4630 Bochum-Querenberg, Germany

The identification of 1-aminocyclopropane-1-carboxylic acid (ACC) as the immediate precursor of ethylene in tissues of higher plants (1,2) has been a major step forward in the elucidation of the biosynthetic pathway, and its regulation, of this plant growth regulator. While ACC-synthase (3,4) appears to control the rate-limiting step in ethylene synthesis in many cases (5,6), an additional control at the conversion of ACC to ethylene has also been observed (7,8). The regulatory pattern of ethylene synthesis appears to be more complex, however, since it has been shown that endogenous, as well as exogenously administered ACC is conjugated with malonic acid to form N-malonyl-ACC (MACC) (9-12). Thus, the rate of formation and hydrolysis of MACC might be an additional factor in the control of the ACC level in a tissue. While a plethora of metabolites and conjugates is known for the other endogenous plant growth regulators and much uncertainty exists about their function in plants (13), it appears advantageous to study the physiology and biochemistry of plant hormone (precursor) conjugation in a less complex system. This paper will describe results of recent experiments on the *in vivo* formation of MACC in various plant tissues and undifferentiated plant cells, as well as on the malonylation of ACC in cell-free extracts of mung bean hypocotyls.

MATERIALS AND METHODS

Tobacco callus was grown on a modified B5 medium containing 0.2 % N-Z-Amine Type A (14) and fortified with 0.8 % Bacto agar. Mung bean *(Vigna radiata* (L.) Wilczek) seedlings were grown in intermittent light (16 h) or darkness, as indicated. For the

Y. Fuchs and E. Chalutz (eds.) Ethylene: Biochemical, Physiological and Applied Aspects.
ISBN 90-247-2984-X. Printed in The Netherlands
© 1984, Martinus Nijhoff/Dr W. Junk Publishers, The Hague.

wilting experiments, 8 cm segments from primary leaves of 9-day-old light-grown wheat *(Triticum aestivum* L.cv. Kanzler) seedlings were processed as described in (8). Conditions for feeding experiments with $[2,3-{}^{14}C]$-ACC, extraction and assay of ACC, MACC, and ethylene, as well as for the identification of MACC by mass spectroscopy are given in (9). Mass-spectroscopic identification of ACC followed the procedure given in (15). For ACC malonyl-transferase extraction, hypocotyl segments from four-day-old light grown mung bean seedlings were ground with pestle and mortar in 0.1 M potassium phosphate buffer, pH 8.0, containing Polyclar AT and sand. Protein in the 39.000 xg supernatant precipitating between 30 and 60 % $(NH_4)_2SO_4$ saturation was used as the transferase source. The assay mixture contained 40 mM potassium phosphate buffer, pH 8.0, 100 mM KCl, 2 mg ml^{-1} bovine serum albumin, 0.5 mM malonyl-CoA, 0.2 mM $[2,3-{}^{14}C]$-ACC (sp.A. 37 mBq mmol^{-1}), and 0.4 mg protein in a total volume of 0.25 ml. ACC and MACC were separated from each other on Dowex WX8(H$^+$-Form), and the radioactivity of the MACC-fraction determined by liquid scintillation counting.

RESULTS

ACC and MACC in tobacco callus.

We had previously reported (10) that certain callus cultures of higher plants contain levels of ACC and/or MACC much higher than those usually found in differentiated plant tissues. Since then it has been reported (16) that concentrations of ACC identified by cochromatography and NaOCl degradation (17) in non-shoot-forming callus of *N.tabacum* exceed 300 nmol g^{-1} FW. We decided to reinvestigate the occurrence of ACC, as well as of MACC, in callus using mass-spectroscopic identification according to (15) and (9), respectively. Fig.1 shows the identification of ACC in callus of *N.alata*, which contains nearly 2 µmol g^{-1} FW ACC as determined by the assay of Lizada and Yang (17), representing ca. 16 % of the total soluble amino acid pool. In Fig.2 the mass spectra of permethylated authentic MACC, and of a highly purified substance from *N.sylvestris* callus are shown.

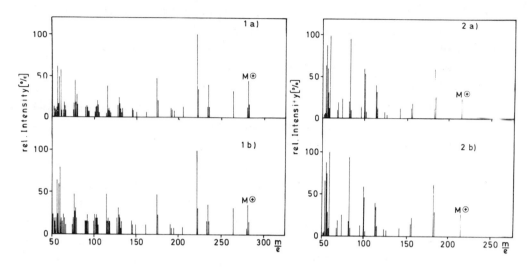

FIGURE 1. Mass spectra of the methylesters of the 2,4-dinitro-
phenyl derivatives of authentic ACC (a) and of
substance isolated from *N.alata* callus (b).

FIGURE 2. Mass spectra of the dimethylesters of authentic MACC
(a) and of substance isolated from *N.sylvestris*
callus (b).

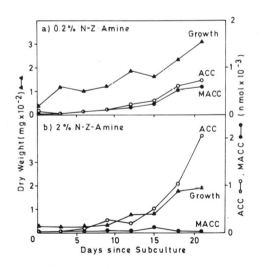

FIGURE 3.

Dry weight increase, and
ACC and MACC content during
growth of *N.tabacum* callus
on medium supplemented with
0.2 % (a) or 2 % (b) N-Z-Amine.
Total amounts per 50 ml culture flask containing 30 ml
solidified medium are plotted
as a function of time.

Again, the identity of the two compounds is evident. The ACC content of this callus was 283 nmol g^{-1} FW, while the MACC content was 658 nmol g^{-1} FW. A callus of *N.tabacum* strain 24-356 contained relatively low amounts of ACC and MACC in roughly equimolar concentrations (ACC: 50 nmol g^{-1} FW; MACC: 40 nmol g^{-1} FW). This ratio did not change significantly during growth on the normal medium (Fig.3a). When, however, the concentration of N-Z-Amine (a pancreatic casein hydrolysate) was increased 10-fold, the cell contained large amounts of ACC, while their MACC content was negligible (Fig.3b). A more detailed analysis revealed that no single amino acid is responsible for this change, but rather certain groups of amino acids. Further work on the regulation of MACC formation in this culture, with particular emphasis on the activity of ACC malonyl transferase (see 3.3.) is currently in progress.

ACC metabolism in water-stressed leaves.

In excised wheat leaves, water-stress results in a transient increase in the rate of ethylene production and the level of ACC, while the level of MACC increases substantially until it reaches a plateau (Fig.4). A renewed water-stress initiates the same series of events once more, as can also be seen from Fig.4. The second transient increase in the ACC level and ethylene release occur with a concomitant <u>increase</u> in the level of MACC, which means that the MACC already present in the tissue is not available for the release of ACC. Thus, MACC must be considered a metabolically stable end product in this system. Results, in essential agreement with these findings, have recently been reported (18). with the exception that benzyladenine had to be supplied during the rehydration period in order to produce another increase in ACC, ethylene, and MACC in response to a second wilting treatment.

ACC malonylation in homogenates of mung bean hypocotyls.

A cell-free preparation from mung bean hypocotyls was found to catalyze the formation of MACC in the following reaction: ACC + malonyl-CoA →MACC + CoA. Some characteristics of this ACC malonyltransferase activity are listed in Table 1.

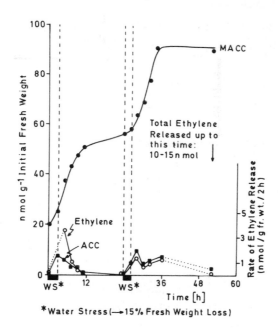

FIGURE 4. Time course of the rate of ethylene release, and
the ACC and MACC content of excised wheat leaves
during two consecutive wilting and rehydration
treatments.

Table 1. Characteristics of malonyl-CoA: ACC malonyltransferase activity from mung bean hypocotyls

pH - Optimum	8 - 8.5
Temperature Optimum	40°C
Stimulation by Anions (0.1 M):	$F^-<SO_4^{2-}<Cl^-<J^-<Br^-<NO_3^-$
app. K_m ACC	0.17 mM
app. K_m Malonyl-CoA	0.25 mM
app. K_i CoA[a]	0.30 mM
app. K_i D-Phenylalanine[b]	1.2 mM

[a] inhibition uncompetitive with respect to malonyl-CoA

[b] inhibition competitive with respect to ACC

Of particular interest is the inhibition of the ACC malonylation by the D-enantiomers of certain nonpolar amino acids, such
as phenylalanine (Table 1), alanine and methionine, because it
has long been known that angiosperm tissues convert D-amino acids
into their N-malonyl conjugates (19). Our preparations from mung
bean hypocotyls malonylate D-phenylalanine and D-alanine, and ACC
serves as an inhibitor in these reactions. Thus, it seems probable that the activities responsible for the malonylation of ACC
and the D-amino acids reside in the same enzyme. Further support
for an intimate relationship between the malonylation of ACC and
D-amino acids comes from *in vivo* studies, in which the inhibitory
activity of D-amino acids on MACC formation from exogenous ACC
was shown to be correlated with their stimulatory activity on
ethylene release (Fig.5). These results, and the results obtained
with the cell-free system fully confirm, complement, and extend
recently published data on the relationship between .the *in vivo*
malonylation of ACC and D-amino acids in mung bean hypocotyls
(20), and they suggest malonylation of ACC as a regulatory factor
in ethylene synthesis.

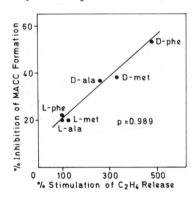

FIGURE 5.

Correlation between inhibition
of MACC formation and stimulation of ethylene release from
exogenously administered
$[2,3-^{14}C]$-ACC in mung bean
hypocotyl sections by the
enantiomers of alanine, methionine, and phenylalanine.

Relationship between malonyltransferase activity and
2-methyl- and 2-ethyl-substituted ACC-analogues.

The stereoisomers of 1-amino-2-ethylcyclopropane-1-carb-
oxylic acid (AEC) have been instrumental in the demonstration
that ACC is stereospecifically converted to ethylene by plant
tissues, since it was shown that only the (1R,2S)-isomer of

AEC is converted to butene (21). With only the *cis*-(1S,2S/1R,2R) and *trans*-(1R,2S/1S,2R)-diastereoisomers of AEC and its methyl homologue (AMC) available to us, we could show that all four compounds are potent inhibitors of ACC malonylation in homogenates of mung bean hypocotyls (I_{50}-values: 0.14 mM *cis*-AMC; 0.07 mM *trans*-AMC; 0.12 mM *cis*-AEC; 0.05 mM *trans*-AEC). Unfortunately, due to the low malonyltransferase activity in the homogenates, we could not determine, whether the ACC-analogues served as substrates of the enzyme, and due to the limited amounts of the compounds available to us, further kinetic analyses of their effect on ACC malonylation were not possible. The following facts were, however, established in *in vivo* experiments (Table 2):

Table 2. Uptake and metabolism of the diastereomers of AMC and AEC by mung bean hypocotyl segments. Ten, 2 cm long, sections from hypocotyls of 3.5-day-old etiolated mung bean seedlings were incubated for 16 h in 0.3 ml of 50 mM potassium phosphate buffer, pH 6.8, 2 % sucrose, 50 µg ml^{-1} chloramphenicol and 3 mM analogue concentration. Analyses were then carried out analogous to ACC feeding experiments (9,10). Results from two independent experiments are given.

Analogue	% of analogue taken up by tissue	% of analogue (taken up by tissue) recovered in:	
		conjugate	butene[a]/propene[b]
trans-AEC	22.7;22.1	0 0	6.6; 2.7[a]
cis-AEC	49.9;49.0	19.1; 4.7	0 0
trans-AMC	39.8;34.0	3.7; 5.4	9.1;10.6[b]
cis-AMC	39.9;35.8	40.8;48.0	4.4; 5.3[b]

It was confirmed that *trans*-AEC, but not *cis*-AEC is converted to 1-butene by mung bean hypocotyls. This is in agreement with data presented in (21). However, *cis*-AMC was converted to 1-propene with 50 % of the efficiency with which *trans*-AMC was converted to 1-propene, indicating a less stringent specificity of the ethylene forming enzyme with AMC-stereoisomers. Extracts of hypocotyls which had been fed

the diastereoisomers of AMC or AEC were subjected to NaOCl
degradation (17) before and after acid hydrolysis. With the exception of *trans*-AEC, the ACC analogues had been conjugated by
the plant tissue, *cis*-AMC being by far the preferred substrate
for conjugation. It is tempting to speculate that the malonyl
conjugate of *cis*-AMC had been formed, but this remains to be
proven. Somewhat disturbingly, *trans*-AEC reduced the incorporation of exogenously administered $[2,3-^{14}C]$-ACC into MACC, while
the other three compounds had no significant effect (data not
shown). Clearly, these experiments should be repeated with the
four separate stereoisomers of each ACC-analogue.

Regulation of ACC malonyltransferase activity.

A variety of (stress)treatments causes a rapid increase in
the rates of synthesis of ACC, ethylene, and also of MACC in many
plant tissues (10,18), and changes in the activity of ACC synthase have primarily been made responsible for the changes in the
rate of ethylene synthesis (5,6,22). Most plant tissues, when
administered ACC, readily malonylate the compound which indicates
that the malonyltransferase is a constitutive enzyme. We have
observed a slight increase (doubling within 6 to 8 h) in the
extractable ACC malonyltransferase activity in mung bean hypocotyl sections in response to 1 mM IAA, but no changes in response
to treatment with a noxious chemical (0.1 mM $CdCl_2$) or 0.5 mM ACC.
Thus, the ACC malonyltransferase, indeed, appears to be a constitutive enzyme.

CONCLUSIONS

The N-malonyl conjugate of ACC - MACC - has now been unequiv-
ocally identified by mass spectroscopy in buckwheat hypocotyls
(9), wheat leaves (11), germinating peanut seeds (12), and callus
of *N.sylvestris* (this paper). No other conjugate of ACC has been
found so far in higher plants. The widespread occurrence of a
mechanism in higher plants by which D-amino acids are conjugated
with malonic acid (19,23), and the apparent relationship between
the malonylation of ACC and D-amino acids in mung bean hypocotyls,
both *in vivo* and *in vitro* (20; this paper; Kionka and Amrhein,

submitted), make it probable that D-amino acids and ACC are malonylated by the same enzyme. While the ethylene forming enzyme is supposed to interact with ACC in a configuration, which corresponds to a L-amino acid (21), the malonyltransferase appears to interact with ACC in a configuration corresponding to a D-amino acid. While the results in Table 2 indicate that the stereospecificity of the ethylene forming enzyme may not be as strict as previously assumed (21), further studies with 2-alkyl-substituted ACC stereoisomers are clearly required to investigate the stereospecificity of the malonyltransferase activity which we have identified in homogenates of mung bean hypocotyls. Our studies, and other published work (12,18), indicate that MACC is a metabolically stable end product - rather than a transiently stored storage compound - which is formed by a presumably constitutive enzyme. In callus tissue, MACC formation can be suppressed by appropriate culture conditions (Fig.3). In mung bean hypocotyls, certain D-amino acids stimulate the formation of ethylene from exogenously administered ACC at the expense of MACC formation (Fig.5 and (20)). Thus, malonylation of ACC may, at least in part, be a factor in the control of ethylene synthesis.

ACKNOWLEDGEMENTS

This work was supported in part by the Minister für Wissenschaft und Forschung des Landes Nordrhein-Westfalen. We thank Mr.Leo Andert for the synthesis of labelled ACC and for recording the mass spectra. We gratefully acknowledge the gift of the ACC-analogues by Dr. M.A. Venis, Shell Research Ltd., Sittingbourne, U.K., and of the callus strains by Professor M.H.Zenk, University of Munich.

REFERENCES

1. Adams DO, Yang SF. 1979. Proc.Natl.Acad.Sci.USA 76,170-174.
2. Lürssen K, Naumann K, Schröder R. 1979. Z.Pflanzenphysiol. 92, 285-294.
3. Boller T, Herner RC, Kende H. 1979. Planta 145, 293-303.
4. Yu YB, Adams DO, Yang SF. 1979. Arch.Biochem.Biophys.198, 280-286.
5. Yang SF. 1980. Hort.Sci. 15, 238-243.
6. Boller T, Kende K. 1980. Nature 286, 259-260.
7. DeLaat AMM, VanLoon LC. 1982. Plant Physiol.69, 240-245.
8. McKeon TA, Hoffman NE, Yang SF. 1982. Planta 155, 437-443.
9. Amrhein N, Schneebeck D, Skorupka H, Tophof S, Stöckigt J. 1981. Naturwissenschaften 68, 619-620.
10. Amrhein N, Breuing F, Eberle J, Skorupka H, Tophof S. 1982. In: Plant growth substances 1982, pp.249-258 (Wareing PF ed.) London, New York, Academic Press.
11. Hoffman NE, Yang SF, McKeon T. 1982. Biochem.Biophys.Res. Commun. 104, 765-770.
12. Hoffman NE, Fu JR, Yang SF. 1983. Plant Physiol.71, 197-199.
13. Sembdner G, Gross D, Liebisch HW, Schneider G. 1980. In: Encycl.Plant Physiol. New Series Vol.9, pp.281-444 (MacMillan J ed.) Berlin, Heidelberg, New York, Springer-Verlag.
14. Zenk MH, El-Shagi H, Schulte U. 1975. Planta medica Suppl., 79-101.
15. Savidge RA, Mutumba GMC, Heald JK, Wareing PF.1983. Plant Physiol. 71, 434-436.
16. Grady KL, Bassham JA. 1982. Plant Physiol. 70, 919-921.
17. Lizada MCC, Yang SF. 1979. Anal.Biochem.100, 140-145.
18. Hoffman NE, Liu Y, Yang SF. 1983. Planta 157, 518-528.
19. Pokorny M, Marčenko E, Keglevič. 1970. Phytochemistry 9, 2175-2188
20. Liu Y, Hoffman NE, Yang SF. 1983. Planta 158, 437-441.
21. Hoffman NE, Yang SF, Ichihara A, Sakamura S. 1982. Plant Physiol. 70, 195-199.
22. Imasaki H, Yoshii H, Todaka I. 1982.See ref.10.pp.259-268.
23. Kawasaki Y, Ogawa T, Sasaoka K. 1982. Agric.Biol.Chem. 46, 1-5.

MODEL SYSTEMS FOR THE FORMATION OF ETHYLENE FROM 1-AMINOCYCLOPROPANE-1-CARBOXYLIC ACID

J.E. THOMPSON, R.L. LEGGE, D.G. McRAE and P.S. COVELLO

Department of Biology, University of Waterloo
Waterloo, Ontario, Canada   N2L 3G1

INTRODUCTION

The terminal step in the ethylene biosynthetic pathway, the conversion of 1-aminocyclopropane-1-carboxylic acid (ACC) to ethylene, exhibits substrate stereospecificity (1) and appears to be membrane-associated (2). As well, studies with protoplasts from Ipomoea flower buds (3) and leaf discs of bean (4) indicate that the reaction is oxygen-dependent and sensitive to radical scavengers. In the present study, we describe two model systems for converting ACC to ethylene -- a chemical system in which the reaction is mediated by the hydroxyl radical and a microsomal membrane system in which the reaction is enzymatic and facilitated by the superoxide radical. These model systems have features in common with the conversion of ACC to ethylene in situ and have provided supporting evidence for the involvement of free radicals in the terminal step in ethylene biosynthesis.

MATERIALS AND METHODS

Membrane isolation and ethylene measurements. Microsomal membranes were isolated from epicotyls of etiolated pea and carnation flowers as described (5,6). Ethylene was measured after a 1-h collection period using a Perkin-Elmer Series 900 gas chromatograph equipped with a flame ionization detector and an alumina column (60 x 0.32 cm). The instrument was operated isothermally at 70°C.

Electron spin resonance spectroscopy. ESR spectra were recorded at room temperature or at 29°C using a Varian E-12 spectrometer equipped with a Varian variable temperature accessory. Aliquots of the reaction mixture were placed in 100 μl capillary tubes sealed at one end and inserted into a quartz sample holder in the microwave cavity. A copper-constantan thermocouple was placed beside the

Y. Fuchs and E. Chalutz (eds.) Ethylene: Biochemical, Physiological and Applied Aspects.
ISBN 90-247-2984-X. Printed in The Netherlands
©1984, Martinus Nijhoff/Dr W. Junk Publishers, The Hague.

capillary tube in the quartz sample holder. g-Values were determined using α,α-diphenol-β-picryl hydrazyl as a standard for which g = 2.0037.

^{13}C NMR spectroscopy. Samples containing 1.0 M ACC and 1.0 M total carbonate in 2 ml were prepared at various pH values using the appropriate amounts of $NaHCO_3$ and Na_2CO_3 and, when necessary, HCl or NaOH. The samples were transferred to 10 mm O.D. NMR tubes with a 4 mm tube containing 99% D_2O fitted coaxially to give the required deuterium lock signal. ^{13}C NMR spectra were recorded using a Bruker WP-80 FT-NMR spectrometer operating at 20.1 MHz with broad band proton decoupling. Samples were first equilibrated for 2 h at 31°C after which 3,000-5,000 free induction decays were accumulated with a recycle time of 12 s at the same temperature. pH values were measured at 31°C within 12 h of recording the spectra.

RESULTS AND DISCUSSION

Free radical-mediated formation of ethylene from ACC. ACC is readily converted to ethylene in the presence of free radicals gene-rated by the xanthine/xanthine oxidase reaction (Fig. 1A) and to a much greater extent when H_2O_2 (33 μM) is included in the reaction mixture (Fig. 1B). As well, catalase (5 units per ml) completely blocked the formation of ethylene (Fig. 1C) indicating that the con-version of ACC to ethylene is dependent upon H_2O_2. The xanthine/ xanthine oxidase mixture produces $O_2^{\bar{}}$, which can spontaneously dispro-portionate to form H_2O_2 (7) and, through Haber-Weiss chemistry (8), the hydroxyl radical (OH$^{\cdot}$). Thus, the inhibitory effect of catalase can be interpreted as indicating that the conversion of ACC to ethylene in the presence of xanthine/xanthine oxidase is mediated by the hydroxyl radical. This is further supported by the fact that ACC is also converted to ethylene in the presence of the Fenton reaction (Figs. 1D&E), which forms the hydroxyl radical directly. Spin-trapping studies with 5,5-dimethyl-1-pyrroline-1-oxide (DMPO) have indicated that the hydroxyl-DMPO spin adduct is detectable in both the xanthine/xanthine oxidase reaction and the Fenton reaction (9). Our inability to detect the superoxide adduct of DMPO as well in the xanthine/xanthine oxidase system presumably reflects the instability of this spin adduct at high pH values and its propensity to spontaneously convert to the corresponding hydroxyl adduct (10). It is to be noted as well that the enhancement of ethylene production achieved by the addition of H_2O_2 to the xanthine/xanthine oxidase system is not attributable to the Fenton reaction inasmuch as stimulation by added H_2O_2 is only achieved when xanthine oxidase is present (9).

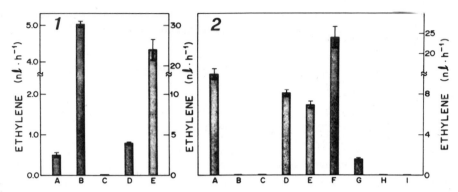

Figure 1. Effects of H_2O_2 and catalase on ethylene production from ACC by the xanthine/xanthine oxidase system and the Fenton reagent. A, 0.2 mM xanthine, 0.05 units xanthine oxidase and 1 mM ACC in 1 ml 2 mM EPPS buffer, pH 8.0; B, A + 33 µM H_2O_2; C, A + 5 units catalase; D, 5 µM FeSO$_4$, 1 mM ACC and 330 µM H_2O_2 in 1 ml 2 mM EPPS buffer, pH 8.0; E, 5 µM FeSO$_4$, 1 mM ACC and 3.3 mM H_2O_2 in 1 ml 2 mM EPPS buffer, pH 8.0. Left ordinate is for A and B; right ordinate is for C, D and E. Standard errors of the means are indicated; n=3. (From reference 9).

Figure 2. Ethylene production from ACC by microsomal membranes from pea epicotyl sections. A, microsomal membranes (200 µg protein/ml), 1 mM ACC in 1 ml 50 mM EPPS buffer, pH 8.0; B, same as A except the microsomal membranes were heat-denatured; C, same as A except anaerobic; D, A + 1 mM H_2O_2; E, A + 5 units catalase; F, A + 8500 units lipoxygenase + 400 µM linolenic acid; G, A + 40 units glutathione peroxidase + 0.08 mM glutathione; H, A + 1 mM n-propyl gallate; I, A + 10 mM Tiron. Standard errors of the means are indicated; n=3. (From references 5 and 13).

Formation of ethylene from ACC by microsomal membranes. Microsomal membranes from a variety of tissues have been shown capable of catalyzing the enzymatic conversion of ACC to ethylene (5,6,11). In the present study, microsomal membranes from pea epicotyls have been used as a model system to further characterize the putative involvement of free radicals in the formation of ethylene from ACC.

The conversion of ACC to ethylene by microsomal membranes from pea proved to be heat-denaturable (Fig. 2A & B) indicating that it is enzymatically mediated, and was completely prevented by the removal of oxygen (Fig. 2C). Like the chemically-mediated reaction (Fig. 1), ethylene formation by microsomal membranes was inhibited by catalase (Fig. 2E). However, the microsomal conversion was also inhibited by H_2O_2 (Fig. 2D), thereby raising the possibility that other hydroperoxides (e.g. lipid peroxides) may be involved. This contention is supported by the finding that the conversion of ACC to ethylene by these membranes is stimulated by lipoxygenase and linolenic acid (Fig. 2F), an exogenous source of lipid hydroperoxides, and inhibited in the presence of glutathione peroxidase and glutathione (Fig. 2G), which catalyse the reduction of hydroperoxides (12).

The conversion of ACC to ethylene by pea microsomal membranes is also inhibited by 1 mM n-propyl gallate (Fig. 2H), an observation that implicates the involvement of free radicals, and evidence obtained by electron spin resonance (ESR) spectroscopy indicates that the super-oxide anion (O_2^-) is required for ethylene production in this system. Tiron (1,2-dihydroxybenzene-3,5-disulfonic acid) reacts specifically with O_2^- to form the Tiron semiquinone free radical, which can be detected as a 4-line ESR spectrum, the amplitude of the spectrum being proportional to the steady state level of O_2^- (5). Thus Tiron can be used to quantify the production of O_2^-. The addition of Tiron to the microsomal reaction mixture for ethylene production gives rise to the characteristic 4-line ESR spectrum reflecting the formation of O_2^- (Fig. 3A). Formation of the Tiron radical signal proved to be oxygen-dependent (Figs. 3B&C), eliminated by heat-denaturation (Fig. 3D), completely inhibited by 1 mM n-propyl gallate (Fig. 3E) and sensitive to superoxide dismutase (Fig. 3F). This latter observation confirms that O_2^- is the radical species being detected by Tiron. The Tiron radical signal generated by the microsomal reaction mixture has also been shown to be unaffected by scavengers of singlet oxygen and of hydroxyl radical (5). Thus, unlike the strictly chemical conversion of ACC to ethylene driven by $OH^\cdot$, the enzymatic reaction appears to be mediated by the less reactive O_2^-.

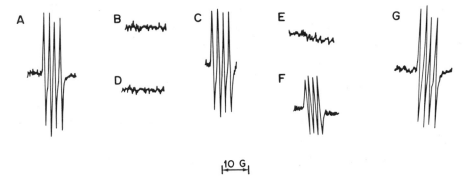

Figure 3. Formation of the Tiron radical signal reflecting O_2^- production by the microsomal reaction mixture capable of converting ACC to ethylene. A, microsomal membranes (200 µg protein/ml), 1 mM ACC, 10 mM Tiron in 1 ml 50 mM EPPS buffer, pH 8.0; B, same as A except anaerobic; C, air was readmitted to the degassed reaction mixture; D, same as A except microsomal membranes were heat denatured; E, A + 1 mM n-propyl gallate; F, A + superoxide dismutase (145 units per ml); G, A minus ACC. (From reference 5).

Tiron completely inhibited the formation of ethylene from ACC by pea microsomes (Fig. 2I). However, formation of the Tiron radical signal by the microsomal reaction mixture was not affected by the omission of ACC (Fig. 3G). This suggests that the superoxide radical acts on ACC, facilitating its conversion to ethylene. Moreover, using the diagnostic spin-trap 4-(N-methylpyridinium) t-butyl nitrone (4-MePyBN), it proved possible to detect an additional free radical in the complete reaction mixture that appears to require ACC, O_2^- and hydroperoxides for its formation. A spectrum of the 4-MePyBN spin adduct of this radical is shown in Fig. 4A. Upon heat denaturation of the microsomal membranes or in the absence of ACC, this radical species was not formed (Fig. 4B&C). Moreover, an essentially similar spectrum was obtained in the chemical system for ethylene production in which the conversion of ACC to ethylene is driven by hydroxyl radicals formed through the Fenton reagent (Fig. 4D). The hyperfine splitting constants for this spectrum are distinctly different from those for the corresponding spin adducts of O_2^- and OH$^\cdot$, indicating that neither superoxide nor the hydroxyl radical is the species being trapped, but they are similar to those for the spin adduct of $^\cdot CH_2OH$ (13). These observations collectively suggest that the radical species detected with 4-MePyBN in the microsomal system for ethylene production is derived from ACC and may well be a carbon-centered radical of ACC. 4-MePyBN also reacts with $O_2^{\overline{\cdot}}$, but the superoxide adduct of this spin trap rapidly disproportionates at pH values of 8.0 and higher (5) and thus does not register in the spectrum recorded for the microsomal system (Fig. 4A).

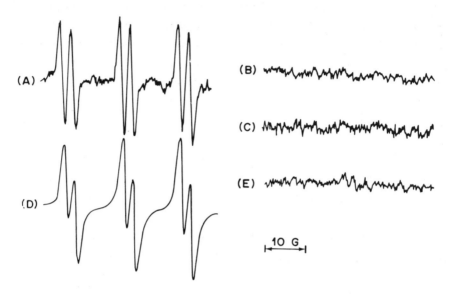

Figure 4. ESR spectra of spin adducts of 4-MePyBN formed in the reaction mixture capable of converting ACC to ethylene. A, microsomes (4.6 mg protein/ml), 0.1 M ACC and 160 mM 4-MePyBN in 50 mM EPPS buffer, pH 8.5; B, A minus ACC; C, same as A except microsomes were heat-denatured; D, 3% H_2O_2, 50 μM $FeSO_4$, 160 mM 4-MePyBN, 0.1 M ACC in 50 mM EPPS buffer, pH 8.5; E, A + 10 mM Tiron. (From reference 13).

Of particular significance is the finding that formation of the 4-MePyBN adduct in the microsomal system proved sensitive to Tiron (Fig. 4E), for this suggests that formation of the ACC-derived radical is dependent upon prior generation of $O_2^{\overline{\cdot}}$. In addition, a 46% reduction in the amplitude of the ACC-derived spectrum was observed when glutathione (0.08 mM) and glutathione peroxidase (40 units per ml) were added, indicating that formation of the ACC-derived radical also requires hydroperoxides. Thus, it would seem reasonable to propose that the ethylene-forming enzyme in this model system mediates the formation of an ACC-derived free radical intermediate prior to ethylene formation in a manner that is dependent upon the presence of $O_2^{\overline{\cdot}}$.

Comparative features of model systems and the native enzyme for converting ACC to ethylene. The model systems have features in common with the native in situ enzyme for converting ACC to ethylene; in particular, they require oxygen and show sensitivity to scavengers of free radicals. However, the efficiency with which the Fenton reaction or the xanthine/xanthine oxidase reaction mediates the conversion of ACC to ethylene is low. For example, the concentration of ACC used in either of these in vitro chemical conversion systems is approximately 1,000-fold higher than endogenous levels found in senescing tissues. This low efficiency presumably reflects the need for steric strain in combination with free radical attack, conditions that would be provided by a substrate-specific enzyme mediating the interaction between ACC and free radicals. The microsomal membrane systems capable of mediating the enzymatic conversion of ACC to ethylene also have a lower conversion efficiency than the native enzyme. Although saturable with ACC, the microsomal enzyme typically exhibits half maximal velocity at an ACC concentration of ≈ 20 mM, which is much higher than the apparent Km value of 0.5 mM ACC reported for protoplasts from senescing flowers of Ipomoea tricolor (3).

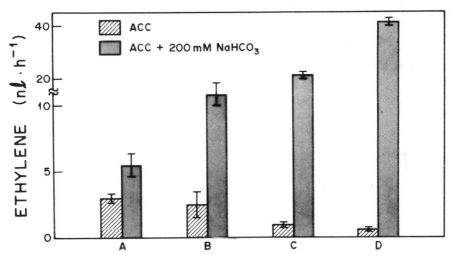

Figure 5. Effects of NaHCO$_3$ on the conversion of ACC to ethylene. A, bean leaf discs; B, carnation flowers; C, microsomes from carnation flowers; D, Fenton reaction. Concentration of ACC was 1 mM in A and D and 2 mM in B and C. (From reference 14).

28

These differences notwithstanding, it is noteworthy that the model systems and the native enzyme for synthesizing ethylene from ACC respond in a similar manner to exogenous $NaHCO_3$. The conversion of ACC to ethylene by bean leaf discs, carnation flowers, microsomal membranes and the Fenton reaction were all markedly enhanced in the presence of 200 mM $NaHCO_3$ (Fig. 5). This effect of $NaHCO_3$ has been attributed to CO_2, and is not due to bicarbonate- or CO_2-induced release of the gas from putative membrane receptor sites (14).

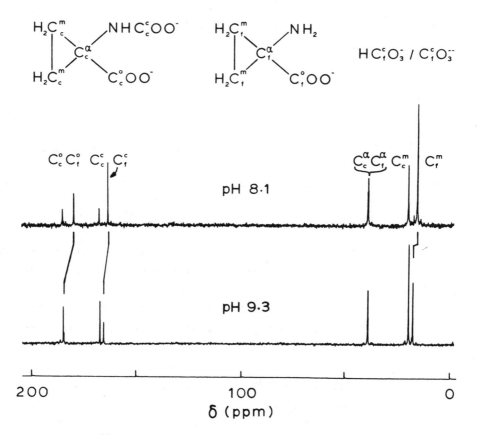

Figure 6. ^{13}C NMR spectra recorded at pH 8.1 and 9.3 for reaction mixtures containing 1 M ACC and 1 M carbonate. The peaks occur in pairs corresponding to the free and carbamate forms of the carboxyl carbon (C_f^o, C_c^o), the carbamate/carbonate carbons (C_f^c, C_c^c), the α carbon (C_f^α, C_c^α -- these are not fully resolved) and the methylene carbons (C_f^m, C_c^m).

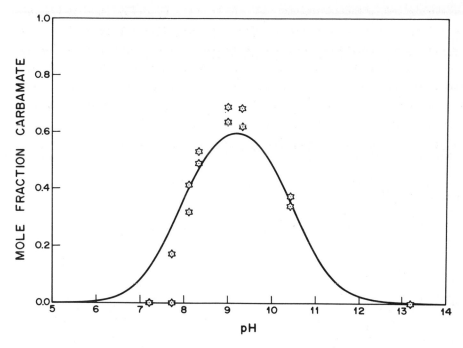

Figure 7. pH dependence of carbamate formation from ACC and carbonate. The data were fitted to the mole fraction equation (15) using a non-linear least squares algorithm.

Since the conversion of ACC to ethylene is enhanced by bicarbonate in chemical systems where no enzyme is present as well as in intact tissue, it is conceivable that CO_2 achieves its effect by interacting directly with ACC, perhaps by forming an ACC-carbamate complex. The ^{13}C NMR spectra illustrated in Fig. 6 confirm that ACC can react with bicarbonate to form an ACC-carbamate complex. The peaks in these spectra occur in pairs corresponding to the various carbon atoms of the free and carbamate forms of ACC. The mole fraction of ACC in the carbamate form (Z) can be calculated using the digitally integrated areas of the peaks for corresponding carbons of the free and carbamate forms of ACC, and by using spectra recorded at different pH values the pH dependence of Z can be determined. By fitting these experimentally determined data points to the mole fraction equation for carbamate formation (15), the equilibrium constant (Kc) for ACC-carbamate was determined. The data show an acceptable fit to the mole fraction equation (Fig. 7), which can be used to predict Z at lower physiological concentrations of ACC and CO_2 by employing the experimentally determined value of Kc (15). Thus, although it has yet to be demonstrated that the carbamate form of ACC is involved in the conversion of ACC to ethylene, it is clear that some endogenous ACC could be expected to be in the form of an ACC-carbamate complex at physiological pH. In this context, it may be significant that the malonyl conjugate of ACC [1-(malonylamino) cyclopropane-1-carboxylic acid], which would be unable to form a carbamate, is not directly converted to ethylene (16).

CONCLUSIONS

Electron spin resonance has provided evidence for involvement of the superoxide anion ($O_2^{\bar{\cdot}}$) radical in the conversion of ACC to ethylene by microsomal membranes. As well, the more reactive hydroxyl radical will mediate the conversion of ACC to ethylene in a strictly chemical reaction that does not involve an enzyme. Although the efficiency of conversion in these systems is lower than that characteristic of senescing tissues, both model systems have in common with the native enzyme the features of being oxygen-dependent, sensitive to free radical scavengers and responsive to $NaHCO_3$. Spin-trap data obtained with the microsomal system have also provided evidence for the formation of an ACC-derived free radical intermediate prior to ethylene formation.

REFERENCES

1. Hoffman NE, Yang SF, Ichihara A, Sakamura S. 1982. Stereospecific conversion of 1-aminocyclopropane-1-carboxylic acid to ethylene by plant tissues. Plant Physiol. 70, 195-199.
2. Apelbaum A, Burgoon AD, Anderson JD, Solomos T, Lieberman M. 1981. Some characteristics of the system converting 1-aminocyclopropane-1-carboxylic acid to ethylene. Plant Physiol. 67, 80-84.
3. Konze JR, Jones JF, Boller T, Kende H. 1980. Effect of 1-aminocyclopropane-1-carboxylic acid on production of ethylene in senescing flowers of Ipomoea tricolor. Plant Physiol. 66, 566-571.
4. Apelbaum A, Wang SY, Burgoon AC, Baker JE, Lieberman M. 1981. Inhibition of the conversion of 1-aminocyclopropane-1-carboxylic acid to ethylene by structural analogues, inhibitors of electron transfer, uncouplers of oxidative phosphorylation and free radical scavengers. Plant Physiol. 67, 74-79.

5. McRae DG, Baker JE, Thompson JE. 1982. Evidence for involvement of the superoxide radical in the conversion of 1-aminocyclopropane-1-carboxylic acid to ethylene by pea microsomal membranes. Plant Cell Physiol. 23, 375-383.
6. Mayak S, Legge RL, Thompson JE. 1981. Ethylene formation from 1-aminocyclopropane-1-carboxylic acid by microsomal membranes from senescing carnation flowers. Planta 153, 49-55.
7. Valentine JS. 1979. The chemical reactivity of superoxide anion in aprotic versus protic media. In Biochemical and Clinical Aspects of Oxygen. Ed. Caughey WS. pp. 659-679. Academic Press, Toronto.
8. Beauchamp C and Fridovich I. 1970. A mechanism for the production of ethylene from methional. The generation of the hydroxyl radical by xanthine oxidase. J. Biol. Chem. 245, 4641-4646.
9. Legge RL, Thompson JE, Baker JE. 1982. Free radical-mediated formation of ethylene from 1-aminocyclopropane-1-carboxylic acid: a spin trap study. Plant Cell Phys. 23, 171-177.
10. Buettner GR, Oberley L. 1978. Considerations in the spin' trapping of superoxide and hydroxyl radical in aqueous systems using 5,5-dimethyl-1-pyrroline-1-oxide. Biochem. Biophys. Res. Commun. 83, 69-74.
11. Mattoo AK, Achilea O, Fuchs Y, Chalutz E. 1982. Membrane association and some characteristics of the ethylene forming enzyme from etiolated pea seedlings. Biochem. Biophys. Res. Commun. 105, 271-278.
12. Little C, Olinescu R, Reid KG, O'Brien PJ. 1970. Properties and regulation of glutathione peroxidase. J. Biol. Chem. 245, 3632-3636.
13. Legge RL, Thompson JE. 1983. Involvement of hydroperoxides and an ACC-derived free radical in the formation of ethylene. Phytochem. 22, 2161-2166.
14. McRae DG, Coker JA, Legge RL, Thompson JE. 1983. Bicarbonate/CO_2 facilitated conversion of 1-aminocyclopropane-1-carboxylic acid to ethylene in model systems and intact tissues. Plant Physiol. 73, 784-790.
15. Morrow JS, Kleim P, Gurd FRN. 1974. CO_2 adducts of certain amino acids, peptides and sperm whale myoglobin studied by carbon 13 and proton nuclear magnetic resonance. J. Biol. Chem. 23, 7484-7494.
16. Hoffman NE, Fu J, Yang SF. 1983. Identification and metabolism of 1-(malonylamino)cyclopropane-1-carboxylic acid in germinating peanut seeds. Plant Physiol. 71, 197-199.

OCCURRENCE OF MEMBRANE-BOUND ENZYME CATALYZING THE FORMATION OF ETHYLENE
FROM 1-AMINOCYCLOPROPANE CARBOXYLIC ACID FROM CARNATION PETALS

SHIMON MAYAK, ZACH ADAM AND AMIHUD BOROCHOV
Department of Ornamental Horticulture, The Hebrew University of Jerusalem,
Rehovot, Israel

INTRODUCTION

Since the identification of 1-aminocyclopropane-1-carboxylic acid (ACC)
as the immediate precursor of ethylene biosynthesis (1,2), the nature of
the reaction converting it to ethylene has drawn considerable attention.
Most of our knowledge of this reaction comes from studies using intact
tissues as model systems (for a review see ref. 3). On the basis of the
findings of these studies, it was suggested that membranes are involved in
the reaction.

RESULTS AND DISCUSSION

Membrane fractions prepared from carnation petals (4) and pea seedlings
(5) have been shown to be capable of catalyzing the reaction converting ACC
to ethylene. Typical results are presented in Table 1, which shows that the
ACC to ethylene conversion activity in carnation petals is associated with
the microsomal membrane fraction and that the activity found in the
supernatant is very low (4).

The membrane system is characterized by a continuously increasing rate
of conversion activity during in vitro incubation. This was explained by
assuming that active enzymes are concealed within the lumina of vesicles
formed during membrane preparation (6), and are thus not immediately
accessible to ACC molecules. Hence the rate of conversion of ACC to

Y. Fuchs and E. Chalutz (eds.) Ethylene: Biochemical, Physiological and Applied Aspects.
ISBN 90-247-2984-X. Printed in The Netherlands
©1984, Martinus Nijhoff/Dr W. Junk Publishers, The Hague.

Table 1. Activity of ACC–dependent ethylene formation in petals and microsomal membranes.

	ACC conversion activity	
	$nl(gfr.wt.)^{-1} h^{-1}$	$nl(mg\ Prot.)^{-1} h^{-1}$
Petals	230	––
Membranes	––	49.8
Supernatant	––	2.7

Membranes were prepared as outlined in (4). Detached petals were enclosed in a test tube and ethylene accumulated in the gas phase was determined. An aliquot of the suspended pellet containing 50 μg protein was added to the reaction mixture consisted of 50 mM EPPS, pH 8.5, 2mM ACC. Incubation was carried at 30 C and the ethylene accumulated during 30 min. was determined.

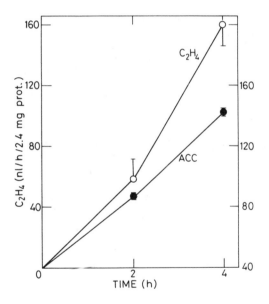

Fig 1. Changes in the rate of ethylene production and ACC content in microsomal membranes from carnation petals as function of time. Membranes were incubated with ACC 2mM, at times indicated the air in the flask was sampled. Then the membranes were centrifuged, the pellet was washed and the ACC content was determined.

ethylene depends on the rate of diffusion of ACC into the vesicles. Apparently the gradual increase in the ACC content of the pellet would be expected to lead to a continuous increase in ethylene production, as indeed occurs (Fig. 1). Further support for the above explanation comes from experiments in which a water-soluble radioactive marker, ^{14}C-sucrose, was incorporated into the vesicles and precipitated with the pellet (Table 2). Following incoporation the pellet was washed and then treated with detergent Nonidet P-40. The results indicated that the marker which previously trapped in the lumen of the vesicles was released into the supernatant. We thus conclude that the rate of conversion of ACC to ethylene by microsomal membranes is influenced by the rate of ACC uptake into the lumen of vesicles formed during the membrane preparation

Rendering the membranes leaky by means of various treatments outlined in Table 3 and then incubating them with ACC, resulted in an increase in the rate of conversion of ACC to ethylene. This could have been due to the fact that the membrane was now more permeable and the active sites thus more accessible to ACC. Of the various treatments used, sonication at low temperature resulted in the greatest increase in conversion activity.

The ACC conversion activity appears to be enzymatic in nature. Most of the supportive data have been published (4), and only the main characteristics of the activity are outlined here as follow:

1. Heat denaturable.

2. TCA and $(NH_4)_2SO_4$ precipitable.

3. Varies with temperature.

4. Varies with pH; optimum pH 8.5

5. Oxygen dependent.

6. Displays saturation with respect to substrate; Km value in the range of 1-2 mM.

Table 2. Release of previously incorporated ^{14}C Sucrose from microsomal membranes by Nonidet P-40.

Fraction	Protein mg	^{14}C —sucrose cpm x 10^3
Before treatment with detergent		
Microsomal-pellet	17.9	338
After Treatment with detergent		
Supernatant	9.0	248
Pellet	7.7	92

Membranes were prepared as outlined in (4), but ^{14}C-sucrose was included in the preparation medium. The pellet was washed three times and the radioactivity was counted. the washed pellet was treated with 0.2% Nonidet P-40, for 1 h, after which the radioactivity in the post-detergent pellet and in the supernatant was counted.

Table 3. Rate of ethylene production by microsomal membranes following freeze-thaw, sonication and Triton X-100 treatments.

Treatment	Ethylene formation %
Control	100
Freeze-thaw	188
Sonication	274
Triton X-100	216

Membranes were prepared and subjected to three cycles of freezing and thawing, or sonication for 1 h at 2 C or treatment with Triton X-100 (0.05% w/v) for 1 h. Aliquots of each of the treated membranes were added to the standard reaction mixture.

The enzyme activity is virtually unaffected by the cations K^{1+}, Ca^{2+} and Cu^{2+}, but is enhanced by Mn^{2+} at an apparent optimum concentration of 5 μM (Fig 2). In the presence of Mn^{2+}, at this optimal concentration, there is a 10-fold increase in specific activity. The enzyme is competitively inhibited by Co^{2+}. with a Ki of 2.3 μM; 50% inhibition is achieved at 7 μM Co^{2+}.

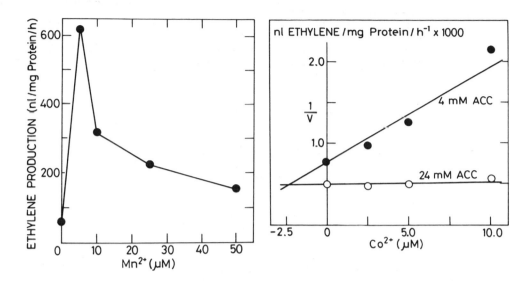

Left: Fig. 2. Effect of Mn^{2+} on ACC conversion activity in the petal membrane fraction.
Various concentrations of MnCl$_2$ were added to a reaction mixture containing 50 mM EPPS pH 8.5, 2 mM ACC and 50 μg of membrane protein. ACC conversion activity was measured as described in Table 1.

Right: Fig. 3. Inhibition of ACC conversion activity in the petal membrane fraction by Co^{2+}.
Various concentrations of CoCl$_2$ were added to a reaction mixture containing 50 mM EPPS pH 8.5, 5 μM MnCl$_2$, 100 ug of membrane protein and either 4 or 24 mM ACC . ACC conversion activity was measured as described in Table 1.

Structural analogs, because of their similar molecular characteristics, might conceivably interfere with substrate—enzyme interactions. If this were indeed the case, one would expect the presence of ACC analogs in the reaction mixture to alter the conversion rate of ACC to ethylene. This possibility was examined by measuring the ACC conversion activity in the presence of an equimolar concentration of each analog. The results presented in Table 4. show that none of the compounds tested was able to inhibit the conversion of ACC to ethylene in intact petals. In the membrane fraction, however the conversion activity was affected (though only slightly) in the presence of two of the analogs, CCA and EI, both of which reduced the conversion of ACC to ethylene by about 15%. Since the analogs were not themselves converted to ethylene, and had little or no inhibitory effect on the conversion of ACC to ethylene, it seems likely that the conversion activity in carnation petals has a strong preference for ACC as substrate.

The specificity of the putative enzyme was further tested by challenging the membrane fraction with various stereoisomers of aminopropylethane carboxylic acid (AEC) as previously done with intact tissue (7). None of the isomers used in our experiments produced 1-butene when incubated with the membrane fraction (Table 5). However, inclusion of allocoronamic acid (a mixture of (1R,2S)-AEC and (1S,2R)-AEC) or coronamic acid (a mixture of (1S,2S)-AEC and (1R,2R)-AEC) in a reaction mixture containing membranes and ACC inhibited ethylene production by 55% and 30% respectively. These results indicate that the putative enzyme interacts specifically with ACC but not with AEC stereoisomers.

Table 4. Effect of ACC analogs on the conversion of ACC to ethylene in intact carnation petal tissues and in a petal membrane fraction.

ACC analog	Ethylene formation	
	Petals $nl(gfr.wt.)^{-1} h^{-1}$	Membrane fraction $nl(mg\ prot.)^{-1}h^{-1}$
—	57 + 5.8	288 + 1.6
AIB	74 + 19.9	294 + 9.1
CCA	53 + 12.5	243 + 3.6
EI	78 + 16.1	245 + 4.8
CA	73 + 8.9	378 + 8.4

Carnation petals were placed in 5 ml vials containing either water (control) or an ACC analog, 2 mM. After 24 h the petals were transferred to a solution containing the analog and 2 mM ACC for a further 24 h, after which they were sealed in 15 ml syringes and ethylene formation was measured. The membrane reaction mixture was similar to the one described in Table 1 except that 2 mM ACC-analog was added. AIB, aminoisobutyric acid; CCA, cyclopropane carboxylic acid; EI, ethyleneimine; CA, cyclopropane-amine. Values are means of 5 replicates +S.E

Table 5. Butene and ethylene formation from ACC and AEC—isomers in a petal membrane fraction.

Substrate	Butene	Ethylene
ACC	*	100%
All AEC—isomers	*	*
ACC + allocoronamic acid	*	45%
" + coronamic acid	*	70%

To the standard reaction mixture 4 mM of ACC or an AEC—isomer or both were added. Ethylene formation was measured as described in Table 1, and butene formation was measured in the same way. (* — not detectable).

Table 6. Effect of different sugars on inhibition of ACC conversion activity by microsomal membranes.

Sugars (40 mM)	Inhibition %
Glucose	70.0
Dextrose	73.0
Sucrose	73.0
Fructose	82.0
Manitol	95.0
Sorbitol	99.5

The physiological relevance of the membrane associated enzyme detected in vitro can be studied by comparing it with the conversion activity studied at the tissue level.

(a) Specificity — both the membrane and the tissue activities show a marked preference for ACC as substrate. When ACC analogs are added, no ethylene is formed. Moreover, the presence of analogs does not inhibit the ACC conversion activity in either system (Table 4).

(b) Inhibitors — both the enzyme and the tissue activities are inhibited by free radical scavengers such as Tiron (8), and propyl gallate (4), indicating that free radicals are required for the reaction to proceed. The fact that sugars have been shown to quench free radicals (9) prompted us to

test the effect of sugars on the ACC conversion activity. Their inhibitory
effect clearly demonstrated in Table 6, and is in agreement with the
findings of Dilley and Carpenter (10) that sugar inhibits ethylene
production by carnation flowers.

Studies on plant tissue have shown that Co^{2+} is involved in inhibiting
the conversion of ACC to ethylene (11). We have shown that Co^{2+} inhibited
the enzyme (Fig 3); 50% inhibition at 7 µM Co^{2+}, it also effectively
inhibited the ACC conversion activity in intact petals by 30% at 10 µM (8).

During the course of preparation of membranes, the occurrence of an as
yet unidentified inhibitor was demonstrated (4). The partial
characterization of the inhibitor is described elsewhere in this volume
(12). The inhibitor influences the activity both in the membrane fraction
and in the intact petal tissue.

(c) The effect of Mn^{2+} in enhancing the ACC conversion activity is
similar in the tissue (1.8–fold increase of activity at 1 µM) and in the
membrane fraction (Fig 2).

On the basis of the numerous similarities between the membrane–bound
enzyme and the ACC conversion activity associated with the tissue, we
suggest that the membrane–bound enzyme described is derived from the in
vivo enzyme system, and is thus of physiological significance. We do not
rule out the possibility that the enzyme may have undergone some structural
alteration or that its localization in relation to other essential
components associated with the membrane may have changed during the course
of membrane preparation. Alterations of this nature might account for the
differences between our own findings and those in other studies, in
particular, the higher Km values (1–2 mM) encountered in our study.
However, it should be pointed out that in using a tissue the possible
active uptake of ACC leading to its accumulation in the tissue cells is

generally neglected. Therefore the ACC concentration values reported in the litrature are those of the outside incubation medium assumed, to reflect the in—cell concentration.

In order to carry out further studies on the enzyme it was necessary to isolate it from the membrane and purify it. Toward achieving this end we have evaluated the use of various detergents. In the presence of ionic detergent the enzymatic activity was abolished. Of the non—ionic detergents tested, Nonidet P—40 was found to be the most suitable. The enzymatic activity was preserved and reasonable solubilization of the enzyme was achieved at an optimum detergent concentration of 0.2% and a protein/detergent ratio of 1:2. When the washed pellet was treated with the detergent, 35% of the enzyme was removed into the supernatant (Table 7). Upon repeating the procedure a total of 57% of the enzyme could be retrieved.

Table 7. Solubilization of the membrane—bound enzyme catalyzing the conversion of ACC to ethylene.

No. of treatments with Nonidet P—40	% solubilization	
	Activity	Protein
x 1	35.0	58.0
x 2	22.1	13.9
Total removed	57.1	71.9

The detergent concentration used was 0.2% and the protein/detergent ratio was 1:2.

The solubilized enzyme was found to resemble the membrane—bound enzyme with respect to optimum pH (8.5), degree of enhancement of activity by Mn^{2+} (12—fold), and degree of inhibition achieved in the presence of 7 μM Co^{2+}.

The solubilized enzyme was further purified on a DEAE — Sepharose column CL6000 (Fig. 4). Upon elution with a linear gradient of KCl, an active fraction centered at fraction no. 8. The specific activity of this fraction was 12-fold higher than that of the solubilized enzyme prepareation placed on the column.

In conclusion we have demonstrated the occurrence of a membrane-bound enzyme capable of catalyzing the conversion of ACC to ethylene, and have shown some of the similarities between this enzyme and the ACC conversion activity studied at the tissue level.

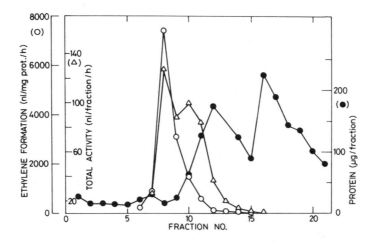

Fig. 4. Elution profile on a DEAE - Sepharose column CL 6000. Solubilized enzyme from microsomal membranes treated with Nonidet P-40, was loaded onto the column. Elution was achieved with a linear gradient of KCl.

REFERENCES

1. Lurssen K, Naumann K, Schroder R. 1979. 1-aminocyclopropane-1-carboxylic acid, an intermediate of ethylene biosynthesis in higher plants. Z. Pflanzenphysiol. 92:285-294.

2. Adams DO, Yang SF. 1979. Ethylene biosynthesis: Identification of 1-amino cyclopropane-1-carboxylic acid as an intermediate in the conversion of methionine to ethylene. Proc. Natl. Acad. Sci. 76:170-174.

3. Yang SF. 1980. Regulation of ethylene biosynthesis. HortScience 15:238–243.

4. Mayak S, Legge RL, Thompson JE. 1981. Ethylene formation from 1–amino cyclopropane–1–carboxylic acid by microsomal membranes from senescing carnation flowers. Planta 153:49–55.

5. Mattoo AK, Achilea O, Fuchs Y, Chalutz E. 1982. Membrane association and some characteristics of the ethylene forming enzyme from etiolated pea seadlings. Biochem. Biophys. Res. Commun. 105:271–278.

6. DeMichelis MI, Pugliarello MC, Rasi–Caldogno F, De Vecchi L. 1981. Osmotic behaviour and permeability properties of vesicles in microsomal preparations from pea internodes. J. Exp. Bot. 32:239–302.

7. Hoffman NE, Yang SF, Ichihara A, Sakamura S. 1982. Stereospecific conversion of 1–aminocyclopropane carboxylic acid to ethylene by plant tissues. Conversion of stereoisomers of 1–amino–2–ethylcyclopropane carboxylic acid to 1–butene. Plant Physiol. 70:195–199.

8. Adam Z, Itzhaki H, Borochov A, Mayak S. 1984. Conversion of 1–amino cyclopropane–1–carboxylic acid to ethylene in intact tissue and in membranes of carnation petals. (submitted).

9. Grimes HD, Perkins KK, Boss WF. 1983. Ozone degrades into hydroxyl radicals under physiological conditions. A spin trapping study. Plant Physiol. 72:1016–1020.

10. Dilley DR, Carpenter WJ. 1975. The role of chemical adjuvants and ethylene synthesis on cut flower longevity. Acta Hortic. 41:117–132.

11. Lau OL, Yang SF. 1976. Inhibition of ethylene production by cobaltous ion. Plant Physiol. 58:114–117.

12. Itzhaki H, Borochov A, Mayak S. 1984. Characterization of an endogenous inhibitor of ethylene biosynthesis in carnation petals. (This volume).

DISTRIBUTION AND PROPERTIES OF ETHYLENE-BINDING COMPONENT FROM PLANT TISSUE

EDWARD C. SISLER
Department of Biochemistry
North Carolina State University
Raleigh, North Carolina 27650-5050, U. S. A.

INTRODUCTION

Ethylene binding ($[^{14}C]$ethylene displaced by unlabeled ethylene) has been observed in a number of plants and plant parts (4, 6). Although proof that the ethylene-binding component is the physiological receptor is lacking, circumstantial evidence points to its being the receptor. The concentration of ethylene necessary to displace $[^{14}C]$ethylene from the binding site is very close to the concentration necessary to give a physiological response. In addition, other compounds which give a physiological response similar to that of ethylene, such as propylene and carbon monoxide, also displace $[^{14}C]$ethylene from the binding site at physiological levels (6).

Studies on the localization of ethylene binding within the cell point to a membrane location (4), probably associated with the endoplasmic reticulum or with protein bodies.

An attempt has been made to purify the binding component from mung bean sprouts, and several hundred-fold purification has been achieved. Many impurities are thought to remain in even the best preparations, and the specific activity remains low.

The amount of binding site in vegetative tissue is low (3.5×10^{-9} mol/kg in tobacco leaves), and 20 or more grams are usually needed for an assay even when $[^{14}C]$ethylene containing in excess of 90% ^{14}C is used. A study was therefore undertaken to determine which plants and plant parts would serve as good sources of the ethylene-binding component for further investigation.

In a previous study, mung bean sprouts were shown to be a convenient source of the binding site (7). Further study showed that dry mung bean seeds had a considerably higher amount of binding component on a weight basis (1); however, when attempts were made to

Y. Fuchs and E. Chalutz (eds.) Ethylene: Biochemical, Physiological and Applied Aspects.
ISBN 90-247-2984-X. Printed in The Netherlands
©1984, Martinus Nijhoff/Dr W. Junk Publishers, The Hague.

purify the component from seeds, some differences were noted between the material from seeds and that from sprouts, raising the possibility that more than one form of the component exists.

This report gives information on the distribution of the binding component and properties of the component from mung bean sprouts and that from seeds.

MATERIALS AND METHODS

Plant material. Seeds were obtained locally. When vegetative material was used, seeds were aerated 6-8 hours and germinated at 25°C in the dark. Mung bean sprouts were obtained from a local market.

Ethylene binding. Ethylene-binding measurements in vivo were made essentially as previously described (6). In vitro measurements were made by mixing the binding component-containing extract with cellulose powder as a spacer and pH 6.0 buffer as previously described (7).

Preparation of plant extracts. Mung bean extracts were prepared and purified as previously described (7), and the fraction eluted from CM-Sephadex was used for isoelectric focusing measurements. When seeds were assayed for binding-site content, they were soaked for 15-20 hours at 4°C and then blended with sufficient 20% Triton X-100 to make the final concentration 2% with respect to detergent. Phosphate buffer at pH 6.0 was included to a final concentration of 0.1 M, and sufficient cellulose powder was then included to make the mixture fluffy to facilitate gas diffusion to the site.

When an extract of the seeds was desired, the soaked seeds were blended with Triton X-100 to a final concentration of 2% and a volume of 2 ml/gram of seeds. After centrifugation at 10,000 x g for 5 minutes, the solution was decanted. The extraction was repeated twice and the extracts were combined and adjusted to pH 4.0. Precipitated material was removed by centrifugation. After adjusting the solution to 35% with $(NH_4)_2SO_4$, it was centrifuged at 10,000 x g for 5 minutes. The clear fluid was siphoned off and discarded. The floating material was cooled to 0°C and treated with acetone at -15°C in a blender. After filtering and drying, the acetone powder was extracted 4 times with 2% Lubrol or Triton X-100. The combined extracts were made to 35% with $(NH_4)_2SO_4$, and the floating material was collected by centrifugation. After dialysis, this material was used for isoelectric focusing.

Isoelectric focusing. This was performed with a preparative flat bed 2117 LKB unit (LKB Instruments Inc.) at 2°C with 5% Ultradex (LKB Instruments Inc.) as support media. Ampholites (pH 4-6 or 6-8) were used at 0.5% with glutamic acid at the anode end and lysine at the cathode end. After an initial 30 minutes at 200 volts, the voltage was raised to 500 for 24 hours. In some experiments, higher voltages were subsequently applied. (In all cases, constant voltage was used.) The extracts containing the binding site were mixed directly with the ampholites and Ultradex (LKB Instruments Inc.) before starting the focusing. After the desired time, the Ultradex was divided into strips, the pH measured, and the sample assayed for binding.

PAGE electrophoresis. This was performed at 700 V for 3.5 hours, essentially by the procedure of Davis (2).

Stability of the binding component to low pH values. A number of samples were adjusted to various pH values and then dialyzed against deionized water. After 6 hours, the samples were removed and assayed for ethylene binding in the standard assay procedure at pH 6.0.

Protein. Protein was estimated by the modified biuret method (5) when Triton X-100 extraction was used and by the absorbance at 260 and 280 nm (9) when a Lubrol extraction was used.

RESULTS

Distribution of ethylene-binding component in seeds. The ethylene-binding component is widely distributed in seeds. The content of a number of seeds is shown in Table 1. It is evident that large differences in the amount of binding component in seeds exist. Of those seeds tested, adzuki beans were the highest, with mung beans containing nearly as much. Most of the legumes had a relatively high amount, but some of the legume seeds were low. Pea (Pisium sativum) had a very low amount. This is of interest in as much as peas have been shown to oxidize ethylene to other products (3). Fava bean (Vicia faba) also has been shown to have a very active oxidation mechanism, although no relationship has been shown between binding and oxidation.

Some of the seeds gave an apparent value of 0 for binding component content; however, they may contain a very small amount. It was necessary to concentrate a Triton X-100 extract from a large amount of seeds to demonstrate the binding component in pea seeds, and no special effort was made with other seeds.

Although most of the other seeds are not as high in binding component as the majority of legume seeds, barley and some others contain a relatively high amount and may serve as a convenient source from a non-leguminous plant for purification studies.

Table 1. Ethylene Binding by Plant Seeds

Plant		Binding	
		dpm/gram	dpm/gram protein[*]
Adzuki bean	Vigna angularis	18,000	80,000
Mung bean	Vigna aureus	16,300	67,650
Field pea	Vigna sinensis	6,870	30,100
Urd	Phaseolus Mungo	6,000	
Cow pea	Vigna sinensis	5,400	23,700
Black bean	Phaseolus vulgaris	4,800	21,300
Black eye pea	Vigna sinensis	4,050	17,750
Pinto bean	Phaseolus vulgaris	3,700	16,700
Soy bean	Glycine Max	3,050	8,900
Bush bean	Phaseolus vulgaris	2,000	9,000
Fava bean	Vicia Faba	600	2,423
Lentel	Lens culinaris	400	1,620
Garbanzo bean	Cicer arietinum	350	
Alfalfa	Medicago sativa	240	
Vetch	Vicia sativa	200	
Pea	Pisum sativum	47	195
Castor bean	Ricinus communis	83	
Cantaloupe	Cucumis Melo	7,530	
Barley	Hordeum vulgare	3,780	39,380
Wheat	Triticum aestivum	1,030	7,370
Oat	Avena sativa	1,300	
Rye	Secale cereale	400	3,330
Watermelon	Citrullus vulgaris	540	
Sunflower	Helianthus annuus	380	1,490
Pumpkin	Cucurbita Pepo	370	1,280
Tobacco	Nicotiana tabacum	320	
Coriander	Coriandrum sativum	330	
Buckwheat	Fagopyrum esculentum	290	2,470
Mustard	Brassica	240	4,020
Cotton	Gossypium hirsutum	230	480
Pop corn	Zea Mays	180	
Millet	Pennisetum	140	1,260
Almond	Prunus Amygdalus	0	
Flax	Linum	0	

[*]Values for protein were obtained from published values (10).

Distribution of ethylene-binding component in vegetative tissue.
Some examples of amounts of ethylene-binding component in vegetative
tissue are presented in Table 2. As in seeds, the amount varies
considerably on a weight basis; however, there is not as much variation
as is found in seeds. It should be noted that pea seeds contain almost
no binding site relative to other legumes, but vegetative tissue, while
low, does contain substantial quantities.

Table 2. Ethylene Binding by Plant Tissues

Plant	Binding dpm/g dry weight		
	Hypocotyl	Leaf	Cotyledons
Adzuki bean	1860	2140	1600
Mung bean	2000	3100	2470
Black eye pea	1600	5080	3600
Soy bean	3680		5700
Bush bean	2300	2700	3700
Watermelon	1440		634
Alfalfa	3500		5300
Pea	624		423

Changes of binding site during germination. Upon imbibition of
water, the amount of binding site declines rapidly after about a day.
As the new tissue develops, the amount of binding site begins to
increase, suggesting that the binding material in the seeds is broken
down and resynthesized.

Extraction of binding material with detergent. Triton X-100 has
been successfully used to extract the binding site; however, it seemed
appropriate to compare its effectiveness with other detergents. Triton
X-100 absorbs light at 280 nm, making direct protein measurements
impossible. Lubrol is nearly as effective as Triton X-100 and did not
interfere with protein measurements (Table 3). Some of the other
detergents were good extractants. The ionic detergent SDS, although not
as effective as some of the non-ionic detergents, extracted substantial
amounts while the quaternary ammonium detergent hexadecyltrimethyl-
ammonium bromide was ineffective.

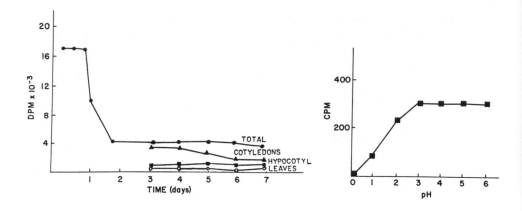

FIGURE 1. Change in ethylene binding by mung beans during germination in darkness. Values are for 1 gram of initial weight.

FIGURE 2. Binding by a Triton X-100 extract of mung bean sprouts after being subjected to various pH values.

Table 3. Effect of Detergent on Extraction of Ethylene-binding Component from Mung Bean Seeds

Detergent	Activity[*] %
Triton X-100	100
Lubrol	93
Brj 30	91
Brj 96	83
Digitonin	81
Sodium lauryl sulfate (SDS)	63
Tween 80	55
Tween 20	43
Sodium cholate	24
Hexadecyltrimethylammonium bromide	0.5

[*]Triton X-100 was arbitrarily taken as 100%.

Stability of the component to low pH. During the extraction and in various purification schemes, low pH conditions are encountered, and it seemed of interest to determine the range within which the component is stable. Fig. 2 shows that the component from sprouts is stable above pH 3.0; however, as the pH is lowered activity is lost, and this could be a crucial factor in some purifcation procedures, particularly isoelectric focusing of the component from seed, since the isoelectric point appears to be low.

PAGE electrophoresis. Initial attempts to purify the ethylene-binding component from mung bean sprouts were by polyacrylamide gel electrophoresis (8). As shown in Figure 3, the material moves as a broad band, and little purifcation could be expected by this method. It does, however, support the idea that the binding component is a membrane protein which does not readily separate from other membrane components.

Isoelectric focusing of mung bean sprout extract. An extract obtained as an eluate from CM-Sephadex (7) was subjected to isoelectric focusing techniques. After initially establishing that the isoelectric point was below pH 7.0, pH 6-8 ampholite was used. The extract contained about 80% of the total activity of a Triton X-100 extract of mung bean sprouts. The remaining 20% of the activity failed to adhere to the CM-Sephadex. After 24 hours at 500 volts, the majority of the binding activity was found in a broad band between pH 5.5-6.5. A small amount appeared below pH 4.0 (Fig. 4) (8). Continued focusing under

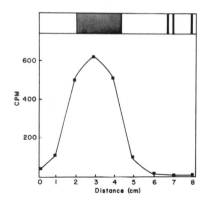

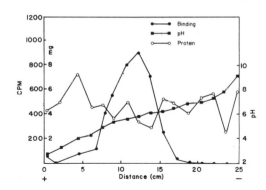

FIGURE 3. PAGE pattern for Triton X-100 extract of mung bean extracts.

FIGURE 4. Isoelectric focusing pattern of Triton X-100 extract from mung bean sprouts. pH 6-8 ampholite.

these conditions did not change the position of the band with respect to pH. If the voltage was raised to 1500 V for a few hours, much of the binding component appeared at a lower pH with a decrease in the amount between pH 5.5-6.5. If focusing was continued for many hours at 1500 volts, the binding component was lost. One of the problems with isoelectric focusing is that the gradient will drift toward the anode end after prolonged periods at high voltages, thus limiting the usefulness of the technique.

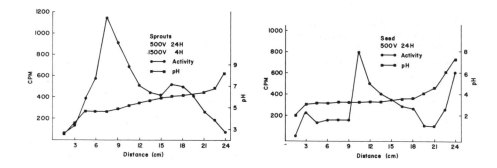

FIGURE 5. Isoelectric focusing pattern of Triton X-100 extract from mung bean sprouts.

FIGURE 6. Isoelectric focusing pattern for a Triton X-100 extract of mung bean seeds. pH 4-6 ampholite.

Isoelectric focusing of an acetone powder of an extract from mung bean seeds. A Triton X-100 extract of mung bean seeds does not adhere to a CM-Sephadex column in a manner similar to a mung bean sprout extract (nor does the original extract before making the acetone powder adhere). When the extract of the acetone powder is subjected to isoelectric focusing using a pH 4-6 ampholite at 500 V for 24 hours, the majority of the component focuses near pH 4.0. However, a small amount appears in the pH 5-7 region, and this apparent low isoelectric point is probably the reason the material did not bind to CM-Sephadex. It is not known whether the binding component obtained from seeds is identical to the component obtained from sprouts or whether it represents a different component. It is possible that the binding component adheres so strongly to other membrane components that 500 volts is not enough to separate it from the other components. Thus, it appears as a broad band which represents a number of components. The location of the band simply represents the area where most of the material moves due to an aggregate isoelectric point. The fact that 1500 V vs. 500 V causes a shift in position toward pH 4.0 would support this hypothesis, and suggests the true isoelectric point may be near pH 4.0.

Table 4. Purification of Ethylene-binding Component from Mung Bean Seeds

Fraction	dpm/g	dpm/mg protein
Triton X-100 extract of seeds	9500	66
Detergent extract of acetone powder from Triton X-100 extract	5128	206

Affinity of the binding component for ethylene. Both of the binding components were tested for their affinity for ethylene under the same conditions. Scatchard plots of the amount of $[^{14}C]$ethylene displaced by unlabeled ethylene from the binding site gave values very close (0.16 µl/l for seeds vs. 0.15 µl/l for hypocotyls and 0.18 µl/l for cotyledons (Table 5). This would add support to the idea that the components from the different plant parts are identical or very similar.

Table 5. Values for the Dissociation Constant (K_d of Ethylene Binding in Mung Beans)

Plant Part	K_d (µl/l)
Seed	0.16
Hypocotyl (7 days)	0.15
Cotyledons (7 days)	0.18

Assayed in the presence of Triton X-100.

DISCUSSION

The results of this study indicate that the ethylene-binding component is widely distributed in plant material; however, the data from seeds indicate that large variations exist. An apparent value of zero in some seeds and 18,000 dpm/g in others raises the question as to what its function would be. The fact that a marked decline in binding component occurs during germination of mung bean seeds may suggest a storage and re-utilization role. Some plants oxidize ethylene to ethylene oxide. Both pea and fava bean, in which oxidation has been studied in some detail (3), are relatively low in the amount of binding component. If the binding component has any relation to the oxidation of ethylene, adzuki beans and mung beans should oxidize ethylene much faster than pea and fava bean. The data presented in this paper should be useful in selecting seeds to test the relation, if any, between binding and oxidation of ethylene.

The data obtained by isoelectric focusing of material from mung bean sprouts illustrates the difficulties in attempting to purify membrane proteins. These proteins interact so strongly that many of the conventional techniques do not work. The broad band of approximately one pH unit for a large portion of the receptor at 500 volts and the shift of this band to a lower value when the voltage is raised to 1500 volts suggests that high voltages might cause a separation of the binding component from other membrane proteins. Recently LKB has introduced ampholites which can be attached to the stationary support. With these it should be possible to use much higher voltages, and perhaps focusing into narrow bands can be achieved.

REFERENCES

1. Beggs MJ, Sisler EC. 1983. Ethylene binding properties of mung bean extracts. Plant Physiol. 72:S226.
2. Davis BJ. 1964. Disc electrophoresis-II. Ann. N. Y. Acad. Sci. 121:404-414.
3. Dodds JH, Hall MA. 1982. Metabolism of ethylene by plants. Int. Rev. Cytol. 76:299-325.
4. Evans DE, Bengochea T, Cairnes AJ, Dodds JH, Hall MA. 1982. Studies on ethylene binding by cell-free preparations from cotyledons of Phaseolus vulgaris L. Subcellular localization. Plant Cell Environ. 5:101-107.
5. Johnson MK. 1978. Variable sensitivity in the microbiuret assay of protein. Anal. Biochem. 86:320-323.
6. Sisler EC. 1979. Measurement of ethylene binding in plant tissue. Plant Physiol. 64:538-542.
7. Sisler EC. 1980. Partial purification of an ethylene-binding component from plant tissue. Plant Physiol. 66:404-406.
8. Sisler EC, Blakistone BA. 1980. Ethylene binding component from plant tissue. Plant Physiol. 65:S43.
9. Warburg O, Christian W. 1941. Isolierung und Kristallisation des Garungsferments Enolase. Biochem. Zeit. 310:384-421.
10. Watt BL, Merrill AL. 1975. Handbook of the nutritional contents of foods. United States Department of Agriculture, Dover Publications Inc, New York.

BINDING SITES FOR ETHYLENE

M.A.HALL, A.R.SMITH, C.J.R.THOMAS & C.J.HOWARTH
Department of Botany & Microbiology, University College of
Wales, Aberystwyth, Dyfed, SY23 3DA, UK.

INTRODUCTION

This article describes recent work on the characterisation
and purification of an ethylene binding site complex (EBS)
derived from developing cotyledons of *Phaseolus vulgaris* L.

MATERIALS AND METHODS

Plant Material

Plants of *Phaseolus vulgaris* L. cv. Canadian Wonder were grown
as described previously (1).

Preparation of extracts and ethylene binding assays

Isolation of the cell-free system, preparation of the solubilised ethylene binding system (EBS), protein determinations
and ethylene binding assay were carried out as in earlier investigations (1,2). Solubilisation of the EBS was also achieved
by suspending the 96,000 x g pellet (2) in 0.02M Tricine, 8%
(w/v) sucrose buffer (pH 7) containing 30mM octyl glucoside.
Ethylene analogues, acetylene and propylene, were prepared
and applied as described previously (3).

To determine the effect of pH on ethylene binding of Triton
X-100 solubilised EBS, $1cm^3$ fractions (equivalent to 1g fresh
weight of cotyledons) were adjusted to give a range of pH
treatments from pH 3 to pH 12. Fractions were then assayed
for ethylene binding activity.

Purification procedures

Gel permeation chromatography was carried out using a Sephacryl column. Prelabelled Triton X-100 solubilised EBS was
applied to a Sephacryl S-300 column (h = 725mm, i.d. = 13mm)
and eluted with 0.02M Tricine, 1% v/v Triton X-100, 0.5M NaCl

Y. Fuchs and E. Chalutz (eds.) Ethylene: Biochemical, Physiological and Applied Aspects.
ISBN 90-247-2984-X. Printed in The Netherlands
©1984, Martinus Nijhoff/Dr W. Junk Publishers, The Hague.

pH 7.   Void volumes of columns were determined using Dextran
2000.

The EBS was partially purified using pH precipitation.   A
series of Triton X-100 solubilised EBS samples were adjusted
over a range pH 3.4-6 and centrifuged on a low speed benchtop
centrifuge for 15min.   The supernatant was decanted and adjusted
to pH 7 and the pellet resuspended in 0.02M Tricine, 0.5% v/v
Triton X-100, pH 7.   Supernatants and resuspended pellets were
assayed for ethylene binding activity and protein.

The procedure for Triton X-114 partitioning was adapted from
Bordier (4).   A Triton X-114 solubilised EBS was heated gently
to $30^{\circ}C$ and centrifuged at 10,000 x g for 30 min, $30^{\circ}C$.   The
lower detergent phase and the upper aqueous phase were collected
and assayed for ethylene binding activity and protein.

Fast protein liquid chromatography (FPLC) was carried out
using a Mono Q anion exchange column. $1cm^3$ (equivalent to 10g
fresh weight of cotyledons) prelabelled with $^{14}C_2H_4$ Triton X-100
solubilised EBS was diluted by 25% in start buffer (0.02M Tris
pH 8.9, 0.5% Triton X-100) and 500μl was loaded onto the column.
The column was eluted with a linear gradient of 0.02M Tris
pH 8.9, 0.5% Triton X-100, 0.35M NaCl (0-100%) at a flow rate
of $2cm^3$ min^{-1} and a pressure of 3MPa.   Absorbance of the eluate
at 280nm was monitored continuously and $0.5cm^3$ fractions
collected which were assayed for ethylene binding activity.

RESULTS AND DISCUSSION

The characteristics of the membrane-bound EBS from *Phaseolus
vulgaris* are well established (1,2).   The K_D of the site for
ethylene is of the order of 10^{-9}-10^{-10}M and its relative affinity for structural analogues of ethylene is broadly similar
to the relative effectiveness of such analogues in mimicking
ethylene effects on growth and development.   The most striking
properties of this site, and of that described by Sisler (5)
in mung beans, are its low rate constants of association and
dissociation (k_1^{25} = $3.18.10^4 M^{-1}s^{-1}$, k_{-1}^{25} = $1.84.10^{-5}s^{-1}$); indeed
it is this feature which has allowed determination of binding
activity *in vivo* and *in vitro* by currently available methods.   The

EBS is present on membranes of the endoplasmic reticulum and protein body membranes (6,7).

We have investigated the properties of both crude and partially purified solubilised preparations. Both the purification and characterisation of the solubilised binding complex have been much complicated by the extreme hydrophobicity of the binding protein which precipitates in the absence of detergent (2).

Of all the treatments we have attempted for solubilising the EBS from membranes, only detergent extraction did not drastically reduce ethylene binding. We have used cholate, Triton X-100 and, more recently, octyl glucoside, although most of the data described below relate to preparations solubilised with Triton X-100. Detergent extraction does not greatly affect the affinity of the EBS for ethylene as shown in Fig.1. Structural analogues of ethylene compete with ethylene for the solubilised EBS (e.g. Fig.2) yielding similar K_i values to those obtained using membrane-bound EBS. The pH optimum for binding in solubilised preparations spans about 3 pH units between pH 4 and pH 7 (Fig.3) this is in contrast to membrane-bound preparations

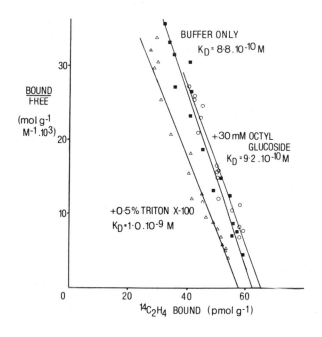

Fig.1. Scatchard plots of ethylene binding to cell-free preparations.

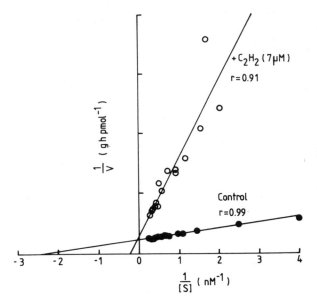

Fig 2. Lineweaver-Burk plots of ethylene binding to Triton X-100 solubilised EBS in the presence and absence of acetylene.

where the optimum is much higher. The sedimentation coeffi-
cient of the EBS-detergent complex derived from isokinetic
sucrose gradients was found to be 2.25S and a Stoke's radius
of 6.1nm was obtained from gel permeation chromatography on
Sephacryl S-300. Calculation of molecular weight by each method

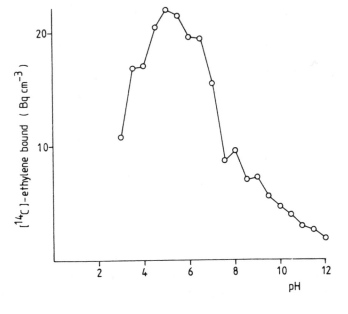

Fig.3. pH depen-
dence of ethylene
binding by Triton
X-100 solubilised
EBS.

leads to values differing by more than an order of magnitude.
However, the frictional ratio was calculated to be 2.4 indicating
that the EBF complex is asymmetric, which accounts for the
disparity in the results obtained by the two methods. Combining
results from both approaches and therefore taking the asymmetry
of the molecule into account yielded a molecular weight of
between 52,000 and 60,000 daltons (8,9). It must be remembered
however that this is the molecular weight of the EBS-detergent
complex and that the value for the EBS protein alone will be
lower.

A summary of the properties of membrane-bound and solubilised
ethylene binding site is given in Table 1.

Table 1.   Comparison of the properties of membrane-bound and
           Triton X-100 solubilised EBS from *Phaseolus vulgaris*

	Membrane-bound preparation	Triton X-100 solubilised preparation
K_D ethylene (M)	0.88×10^{-10} (at infinite site dilution)	5.5×10^{-10}
K_i propylene (M)	5.6×10^{-7}	0.92×10^{-7}
K_i acetylene (M)	1.03×10^{-5}	1.62×10^{-5}
pH optimum	$7.5 - 9.5$	$4 - 7$

We have investigated a number of approaches to the problem of
purifying solubilised EBS, including anion and cation exchange
chromatography, gel permeation chromatography, detergent partitioning, electrophoresis, isoelectric focussing, heparin
affinity chromatography, phase partitioning, pH precipitation
and fast protein liquid chromatography (FPLC). Of these,
enzymic digestion followed by GPC, pH precipitation, detergent
partitioning and FPLC have proved most effective.

Over the range pH3.4-6 most of the protein in solubilised
EBS preparations was precipitated, leaving most of the ethylene
binding activity in the supernatant. This treatment resulted
in an approximately six-fold purification (Fig.4). Equally, a
seven-fold purification was achieved on Sephacryl S-300 when
EBF preparations were pretreated with α-chymotrypsin (Fig.5).

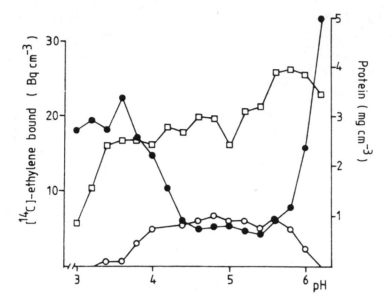

Fig.4. pH precipitation of Triton X-100 solubilised EBS. Binding measured in supernatant (□) and pellet (O). Protein in supernatant (●)

Triton X-114, a detergent which partitions at 30°C into separate detergent and aqueous phases from aqueous solutions of the detergent, was used to solubilise EBS from membranes and

Fig.5.   GPC of α-chymotrypsin pretreated Triton X-100 solubilised EBS on Sephacryl S-300. Protein (O); specific activity of fractions (●); specific activity of loaded extract (■).

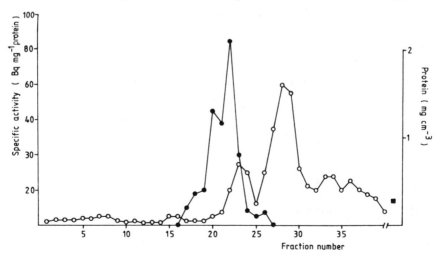

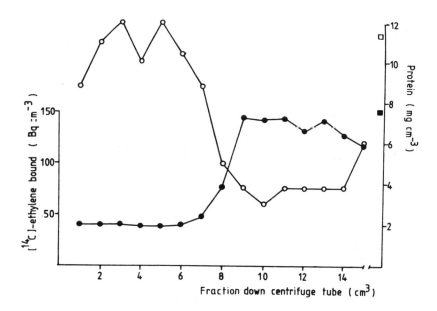

Fig.6. Partitioning of Triton X-114 EBS. Protein (○); ethylene binding activity (●).

the preparation used to determine whether a purification could be obtained on partitioning. As shown in Fig.6 a very marked purification was obtained as most of the protein in the sample

Fig.7. FPLC of Triton X-100 solubilised EBS on a Mono Q column. Protein (solid line); salt concentration (dashed line); binding activity (bars).

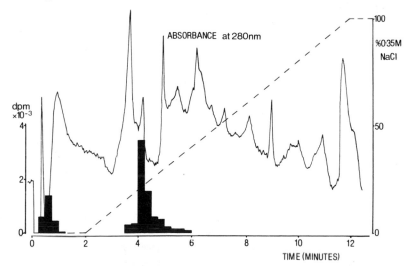

remains in the aqueous phase whereas most of the binding
activity remains in the detergent.

However, by far the best results have been obtained with
FPLC using an anion exchange column. The results from a
typical separation are shown in Fig.7. The method yields
substantial purification and has the added advantage of rapidity
and high recovery.

We are at present involved in subjecting solubilised EBS
to a combination of the techniques outlined above in order to
purify the solubilised EBS to homogeneity. This work has two
ends in view namely a) to subject the EBS protein to analysis
via amino-acid sequencing and protein reactive agents b) to
prepare specific antibodies to the EBS with a view to developing
immunoassay procedures to determine the location and concentration of EBS in parts of the plant (e.g. abscission zones)
known to contain EBS but where its concentration is too low
to permit reliable quantitation by present methods.

ACKNOWLEDGEMENTS

We are grateful to the ARC & SERC for financial support of
this work.

REFERENCES

1. Bengochea, T., Acaster, M.A., Dodds, J.H., Evans, D.E.,
   Jerie, P.H. and Hall, M.A. 1980. Studies on ethylene binding
   by cell-free preparations from cotyledons of *Phaseolus vulgaris*
   L.: Effects of structural analogues of ethylene. Planta *148,*
   397-406.
2. Thomas, C.J.R., Smith, A.R. and Hall, M.A. 1984. The effect
   of solubilisation on the character of an ethylene binding
   site from *Phaseolus vulgaris* L. cotyledons. Planta (in press).
3. Bengochea, T., Dodds, J.H., Evans, D.E., Jerie, P.H., Niepel,
   B., Shaari, A.R. and Hall, M.A. 1980. Studies on ethylene
   binding by cell-free preparations from cotyledons of
   Phaseolus vulgaris L.: Separation and characterisation. Planta
   148, 407-411.
4. Bordier, C. 1981. Phase separation of integral membrane
   proteins in Triton X-114 solution. Journal of Biological
   Chemistry *256,* 1604-1607.
5. Sisler, E.C. 1982. Ethylene binding properties of a Triton
   X-100 extract of Mung Bean sprouts. J. Plant Growth Regulation *1,* 211-218.
6. Evans, D.E., Dodds, J.H., Lloyd, P.C., ap Gwynn, I. and
   Hall, M.A. 1982. A study of the subcellular localisation of an

ethylene binding site in developing cotyledons of *Phaseolus vulgaris* L.: Subcellular localisation. Plant, Cell & Environment *5*, 101-107.

7. Evans, D.E., Dodds, J.H., Lloyd, P.C., ap Gwynn, I. and Hall, M.A. 1982. A study of the subcellular localisation of an ethylene binding site in developing cotyledons of *Phaseolus vulgaris* L. by high resolution autoradiography. Planta *154*, 48-52.

8. Noll, H. 1967. Characterisation of macromolecules by constant velocity sedimentation. Nature *215*, 360-363.

9. Siegel, L.M. & Monty, K.J. 1966. Determination of molecular weights and frictional ratios of protein in impure systems by the use of gel filtration and density gradient centrifugation: Application to crude preparations of sulfite and hydroxylamine reductase. Biochimica et Biophysica Acta *112*, 346-362.

[18]O studies indicate a monooxygenase is involved in ethylene metabolism (12). Based principally on studies in non-biological systems (15,25), copper (Cu^+ but not Cu^{++}) is believed to be involved in ethylene complexation and oxidation. Importantly, not all tissues carry out the extended or complete oxidation pathway. *Vicia faba* cotyledons rapidly convert ethylene to ethylene oxide, but not CO_2 (9). In some tissues, CO_2 is the predominate gaseous product, while in others, both CO_2 and ethylene oxide are produced. These two volatile metabolites are readily recovered in an alkaline trap (ethylene oxide is rapidly hydrolyzed in base and recovered as ethylene glycol). In tissues where both are released, the ratio of CO_2 to ethylene oxide produced varies greatly depending on the tissue, stage of growth, ethylene concentration and exposure time (9,10,11).

A wide variety of different tissues have been demonstrated to metabolize ethylene (Table 1). Alfalfa seedlings and *Vicia* cotyledons have the highest reported rates. The apparent widespread occurrence of this hormone metabolic system in plants, coupled with its apparent high degree of regulation, suggests a possible physiological role.

Table 1. Several Plant Tissues Reported to Metabolize Ethylene

Plant Tissue	Relative Metabolism Rate	Reference
Alfalfa Seedlings (*Medicago sativa*)	+++	Unpublished*
Broad Bean Cotyledons (*Vicia faba*)	+++	18,24
Carnation Flowers (*Dianthus caryophyllus*)	++	4
Pea Seedlings (*Pisum sativum*)	++	2,3,9
Tomato Fruit (*Lycopersicon esculentum*)	+	11
Cotton Abscission Zones (*Gossypium hirsutum*)	+	7
Morning Glory Flowers (*Ipomoea tricolor*)	+	5
Soybean Seedlings (*Glycine max*)	+	Unpublished*
Bush Bean Seedlings (*Phaseolus vulgaris*)	+	"
Wheat Seedlings (*Triticum aestivum*)	+	"

*Beyer, unpublished data

PHYSIOLOGICAL ROLE

Assigning a physiological role to ethylene metabolism is a major challenge facing current and future research in this area. It is counter productive to suggest that ethylene metabolism serves no useful purpose. Not only does this seem highly unlikely, but even if true, other possibilities must first be eliminated. Based on current data, the following three roles for ethylene metabolism in plants need to be carefully examined. First, ethylene metabolism may serve to reduce endogenous ethylene levels; secondly, the oxidation of ethylene at the hormone binding site may be a requirement for ethylene action; and thirdly, the product(s) of ethylene oxidation may serve to alter tissue sensitivity to ethylene. The evidence for and against these three possibilities is briefly reviewed.

Ethylene Removal

There is one recent report which provides some evidence in favor of metabolism as a mechanism for ethylene removal (19). This study examined the relationship between resistance to waterlogging and ethylene metabolism in 21 different cultivars of *Vicia faba*. Some correlation was found to exist between the ability of these cultivars to withstand waterlogging and to metabolize ethylene in the cotyledonary tissue. It was suggested that resistance to waterlogging may be related to the plant's ability to effectively metabolize (remove) the increased amounts of ethylene produced in response to water-logging. Unfortunately, measurements of ethylene oxidation to ethylene oxide were made only in the cotyledons. Thus, further work is needed to determine if the greater capacity of resistant cultivars to metabolize ethylene also extends to roots and shoots where ethylene removal would be especially critical.

There are several reasons for concluding that ethylene metabolism does not represent a system for regulating endo-genous ethylene levels. First, the total amount of ethylene metabolized by plants appears too small to constitute an effective removal system (9,10,11). However, this needs to be carefully examined in those tissues where the rates are exceptionally high, such as in alfalfa seedlings (unpublished data, Beyer) and *Vicia faba* cotyledons (24). Typically, less than 10% of the ethylene produced by plant tissues is destroyed through metabolism. In some tissues such as morning glory flowers, it is only about 0.2% even during periods of maximum production (5). Experiments with CS_2 support the ineffectiveness of metabolism as an ethylene removal system. For example, CS_2 at 10 μℓ/ℓ markedly inhibits ethylene metabolism without affecting ethylene production (4). If metabolism were removing a significant amount of the ethylene being produced, such treatments should result in a detectable increase in the ethylene evolved from the tissue. Since such increases in ethylene evolution following CS_2 treatment are not observed ethylene metabolism probably does not remove a significant amount of ethylene in most tissues.

Secondly, ethylene production rates and periods of peak production do not always correlate well with ethylene metabolism. Such a relationship would be expected if the purpose of ethylene metabolism were to reduce ethylene tissue levels.

Thirdly, ethylene readily diffuses out of most tissues. Thus, regulation of ethylene biosynthesis provides an effective means of controlling ethylene tissue levels making an inactivation system appear unnecessary.

Metabolism - Action Hypothesis

It has been suggested that ethylene metabolism may be required for ethylene to carry out its biological function (2,3,8). The principal reasons against this idea are the following. First, ethylene action and metabolism do not have similar dose response curves. Most ethylene responses saturate at ethylene concentrations between 1 and 10 μℓ/ℓ. In marked contrast, the rate of ethylene metabolism increases fairly linearly up to 100 μℓ/ℓ (3). This means that, if ethylene action and metabolism were related, some step beyond the initial ethylene oxidation step(s) would then have to become rate limiting above 1 to 10 μℓ/ℓ. Alternatively and as already suggested (8), if the ratio of ^{14}C-ethylene metabolites recovered in the tissue to $^{14}CO_2$ produced were critical, this may explain the apparent anomally. This ratio rapidly changes and even reverses over the concentration range of 0.1 to 1.0 µl/l (3).

Secondly, in peas propylene is metabolized much more rapidly than ethylene yet it is about 100 times less effective than ethylene in inducing characteristic ethylene responses (6). Several major differences do exist however between ethylene and propylene metabolism. Most notably, the products of propylene metabolism are mainly three carbon compounds (propylene oxide, propylene glycol and its conjugate) and the distribution of products between the gas phase and tissue is reversed from that observed with ethylene.

Thirdly, while the affinity of the ethylene metabolism system in *Vicia* corresponds well with concentrations required to induce a half-maximal response in the pea-stem growth assay (K_D = 4.2 x 10^{-10} M or 0.1 µl/l in gas phase), the affinity of the metabolism system in peas is several orders of magnitude lower (20,23). This discrepancy is difficult to reconcile unless ethylene metabolism were to alter ethylene tissue sensitivity (see below) or receptor affinity. It is puzzling that the affinity for ethylene metabolism in peas should differ so markedly from that in *Vicia*. Equally puzzling is the lack of the ethylene to CO_2 forming system in *Vicia*.

Countering the evidence against the ethylene metabolism-action hypothesis is a substantial amount of data which suggests that such a relationship might exist. First, and perhaps most noteworthy, is the quantitative relationship that has been demonstrated to exist between the effects of the powerful ethylene antagonist, Ag^+, on ethylene action and metabolism (8). In peas Ag^+ reduced ethylene induced growth inhibition and metabolism (i.e., in terms of ^{14}C-metabolites recovered in the tissue) in a remarkably parallel fashion. The inhibitory effect of Ag^+ on ethylene metabolism was recently confirmed in the same tissue (20). Secondly, the

other antagonist of ethylene action, CO_2, also modifies
ethylene metabolism (8,20). Increasing the ambient CO_2
concentration from 4 to 10% inhibits the metabolism of
ethylene to CO_2 (20). Interestingly, the ethylene metabolites
recovered from the tissue were unaffected by treatments of up
to 5% CO_2. Above this level, however, CO_2 caused a marked
stimulation in the amount of tissue metabolites produced.

Both the CO_2 and Ag^+ studies, indicate that certain aspects
of ethylene metabolism can be inhibited by two of the most
potent and specific inhibitors of ethylene action known. The
significance of the ability of Ag^+ and CO_2 to inhibit different
parts of the ethylene metabolic system is not known but could
explain the greater potency and non-competitive nature of the
Ag^+ effect. The ability of CO_2 to either increase or decrease
certain aspects of ethylene metabolism is particularly
interesting since CO_2 is known to inhibit ethylene action in
some situations and to mimic or enhance it in others (1).

Thirdly, a number of correlations have been found to exist
between changes in tissue sensitivity to ethylene and changes
in ethylene metabolism. Morning glory flower buds which do
not respond to ethylene also have been found to lack the
ability to metabolize ethylene (5). However, just as the
buds become responsive, this ability appears and then
increases rapidly. Similar parallel relationships have been
noted in abscission zone tissue during cotton leaf abscission
(7). A constant rate of ethylene metabolism was observed in
the separation zone tissue prior to abscission induction.
Induction of abscission by deblading resulted in over a six-
fold increase in metabolism and this increase preceded by one
day the first signs of abscission. Hormone treatments that
delayed or stimulated abscission had a parallel effect on
ethylene metabolism.

Altered Tissue Sensitivity

Since ethylene oxide is central to the overall ethylene
oxidation process, its effects on ethylene metabolism and
ethylene action are of considerable interest. In contrast

to earlier work (1) which suggested that ethylene oxide may
be an ethylene antagonist, recent studies (9, unpublished
data, Beyer) strongly suggest that the opposite is true, i.e.,
ethylene oxide increases tissue sensitivity to ethylene.

Ethylene oxide will not mimic ethylene when applied alone.
At high concentrations (>250 µℓ/ℓ), it typically arrests
growth and development. Visual toxic effects are generally
lacking unless very high concentrations are employed (>1000
µℓ/ℓ). Undoubtedly, the earlier reports suggesting that
ethylene oxide blocks ethylene action in flowers and fruits
were due to a general suppression of cellular metabolism
resulting from the high ethylene oxide concentrations employed
(500-7500 µℓ/ℓ). These concentrations tend to "freeze" plant
metabolic functions making normal responses to ethylene
impossible.

Three different systems have been used to demonstrate the
synergistic effect of ethylene oxide, namely the 'triple
response' in peas, leaf abscission in cotton and growth
stimulation in rice. Importantly, ethylene oxide does not
stimulate ethylene production in these tissues. In dark
grown pea seedlings, all of the characteristic effects of
ethylene were enhanced when ethylene oxide (25 to 100 µℓ/ℓ)
was combined with ethylene (0.25 µℓ/ℓ) in a continuous gas
flow system. Similarly, in cotton abscission, ethylene oxide
(150 µℓ/ℓ) in combination with ethylene (0.8 µℓ/ℓ) accelerated
the abscission of debladed cotyledonary petioles. With rice
seedlings, ethylene oxide (4 µℓ/ℓ) was found to significantly
increase the effectiveness of ethylene (7.0 µℓ/ℓ) in stimulating rice coleoptile growth.

These and other similar studies raise the question of
whether or not the products of ethylene metabolism such as
ethylene oxide might modulate tissue sensitivity to ethylene.
Clearly, the levels of ethylene oxide used in these studies
exceed those considered to be physiological based on rates
of ethylene metabolism in most tissues. However, this
possibility cannot be ruled out since ethylene oxide or
other metabolites formed in vivo may be much more effective

than when applied exogenously. It is known for example that
in peas ethylene oxide is not readily converted to ethylene
glycol or CO_2 when applied exogenously, yet _in_ _vivo_ both are
readily formed provided the precursor is ethylene.

BURGS' ACTION HYPOTHESIS

In addition to the various relationships mentioned above,
it should be noted that the scheme proposed by Burg and Burg
in 1965 for ethylene action (14) has certain elements
remarkably similar to the ethylene metabolism-action
hypothesis. Based on their mechanism of action studies,
they proposed that ethylene action involves the binding of
ethylene and oxygen to the metal of a metalloenzyme. Ethylene
was viewed strictly as a dissociable activator molecule. If
this scheme were modified slightly so that ethylene reacted
with the oxygen to form ethylene oxide, thereby initiating
the ethylene oxidation and action sequence, the basic concepts
that have emerged from work on ethylene metabolism and
ethylene action would be reconciled into one scheme.
Interestingly, the enzyme proposed by the Burgs would be a
monooxygenase.

CONCLUSIONS

Ethylene metabolism in plants has several possible
physiological functions. There are data for and against
each possibility. It would be premature at this time to
discard any of them. Ethylene metabolism should be viewed
as a relatively new doorway through which future investigations might shed light on the regulatory nature of ethylene.
This is also true for ethylene binding (10). It is the view
of the author that at some point these two lines of investigation will merge and provide new insight into the mechanism
by which ethylene so profoundly affects plant growth and
development.

REFERENCES

1. Abeles, F. B. (1973). Ethylene in Plant Biology. Academic Press, New York, pp. 302.
2. Beyer, E. M., Jr. (1975). ^{14}C-Ethylene incorporation and metabolism in pea seedlings. Nature 255, pp. 144-147.
3. Beyer, E. M., Jr. (1975). $^{14}C_2H_4$: Its incorporation and metabolism by pea seedlings under aseptic conditions. Plant Physiol. 56, pp. 273-278.
4. Beyer, E. M., Jr. (1977). $^{14}C_2H_4$: Its incorporation and oxidation to $^{14}CO_2$ by cut carnations. Plant Physiol. 60, pp. 203-206.
5. Beyer, E. M., Jr. (1978). $^{14}C_2H_4$ Metabolism in morning glory flowers. Plant Physiol. 61, pp. 896-899.
6. Beyer, E. M., Jr. (1978). Rapid metabolism of propylene pea seedlings. Plant Physiol. 61, pp. 893-895.
7. Beyer, E. M., Jr. (1979). [^{14}C]-Ethylene metabolism during leaf abscission in cotton. Plant Physiol. 64, pp. 971-974.
8. Beyer, E. M., Jr. (1979). Effect of silver ion, carbon dioxide, and oxygen on ethylene action and metabolism. Plant Physiol. 63, pp. 169-173.
9. Beyer, E. M., Jr. (1980). Recent advances in ethylene metabolism. In Aspects and Prospects of Plant Growth Regulators, DPGRG/BPGRG. Monograph 6, pp. 27-38.
10. Beyer, E. M., Jr. (1981). Ethylene action and metabolism. In Recent Advances in the Biochemistry of Fruits and Vegetables, pp. 107-121.
11. Beyer, E. M., Jr. and Blomstrom, D. C. (1980). Ethylene metabolism and its possible physiological role in plants. Proc. Tenth Int. Conf. Plant Growth Subs. (1979), ed. F. Skoog, pp. 208-218. Springer-Verlag, Berlin.
12. Beyer, E. M., Jr., Morgan, P. W. and Yang, S. F. (1984). Ethylene. In Advanced Plant Physiology, ed. M. B. Wilkins, Pitman Books, Ltd., London, pp. 111-126.
13. Blomstrom, D. C. and Beyer, E. M., Jr. (1980). Plants metabolise ethylene to ethylene glycol. Nature 283, pp. 66-68.
14. Burg, S. P. and Burg, E. A. (1965). Ethylene action and the ripening of fruits. Science 148, pp. 1190-1196.
15. Buxton, G. V., Green, J. C. and Sellers, R. M. (1976). J. Chem. Soc., Dalton Trans. pp. 2160-2165.
16. Dodds, J. H. and Hall, M. A. (1982). Metabolism of ethylene by plants. Int. Rev. Cytol. 76, pp. 299-325.
17. Dodds, J. H., Heslop-Harrison, J. S. and Hall, M. A. (1980). Metabolism of ethylene to ethylene oxide by cell-free preparations from Vicia faba L. cotyledons: Effects of structural analogues and of inhibitors. Plant Science Letters 19, pp. 175-180.
18. Dodds, J. H., Musa, S. K., Jerie, P. H. and Hall, M. A. (1979). Metabolism of ethylene to ethylene oxide by cell-free preparations from Vicia faba L. Plant Science Letters, 17, pp. 109-114.
19. Dodds, J. H., Smith, A. R. and Hall, M. A. (1982/83). The metabolism of ethylene to ethylene oxide in cultivars of Vicia faba: Relationship with waterlogging resistance. Plant Growth Regulation 1, pp. 203-207.

20. Evans, D. E., Smith, A. R., Taylor, J. E. and Hall, M. A.
    Ethylene metabolism in *Pisum sativum* L.: Kinetic parameters,
    the effects of propylene, silver and carbon dioxide and
    comparison with other systems (in press).
21. Giaquinta, R. T. and Beyer, E. M., Jr. (1977). $^{14}C_2H_4$:
    Distribution of ^{14}C-labeled tissue metabolites in pea
    seedlings. Plant and Cell Physiol. 18, pp. 141-148.
22. Hall, M. A. (1983). Ethylene receptors. In Receptors in
    Plants and Cellular Slime Moulds, eds. C. M. Chadwick and
    D. R. Garrod, Marcel Dekker, New York (in press).
23. Hall, M. A., Evans, D. E., Smith, A. R., Taylor, J. E. and
    Al-Mutawa, M. M. A. (1982). Ethylene and senscence. In
    Growth Regulators in Plant Senescence, British Plant Growth
    Regulator Group, Monograph 8, pp. 103-111.
24. Jerie, P. H. and Hall, M. A. (1978). The identification
    of ethylene oxide as a major metabolite of ethylene in
    Vicia faba L. Proc. R. Soc. Lond. B. 200, pp. 87-94.
25. Thompson, J. S., Harlow, R. L. and Whitney, J. F. (1983).
    Copper (I)-olefin complexes. Support for the proposed role
    of copper in the ethylene effect in plants. J. Am. Chem.
    Soc. 105, pp. 3522-3527.

ROLE OF ETHYLENE OXIDATION IN THE MECHANISM OF ETHYLENE ACTION

F. B. ABELES

USDA ARS, Appalachian Fruit Research Station, Route 2, Box 45,
Kearneysville, WV 25430, USA

Oxidation of ethylene to ethylene oxide by ethylene monooxygenase
(EM) has been shown to occur in Mycobacterium paraffinicum (14, 16,
17, 18, 32), plants (8, 21, 24), and animals (22, 23). In
microorganisms and animals, ethylene oxide is the only product of the
reaction. In plants however, oxidation of ethylene also yields
CO_2. The ethylene oxide formed in fababeans eventually gives rise
to ethylene glycol, ethanolamine, oxalate, and glycolate (20).

Plants possessing the ability to produce both ethylene oxide and
CO_2 include Pisum sativum (3), Dianthus caryophyllus (5), and
Ipomoea tricolor (10). Vicia faba (fababean), on the other hand,
produced only ethylene oxide (8, 21, 24). In addition, fababean EM is
present in high levels and can be measured by following the uptake of
ethylene or other hydrocarbons from the gas phase surrounding plant
tissue. The data shown in Table 1 indicate that, except for root
tips, all portions of fababean seedlings consume ethylene. It is not
known if root tip tissue actually lacks the oxidative system or
whether the rate of ethylene production exceeds the rate of oxidation.

The fababean system is a favorable one to use in studies dealing
with the role of EM in ethylene action for the following reasons.
First, since only one oxidation pathway occurs in this tissue,
interpretation of data is simplified. Second, since EM activity is
high enough to be measured directly by gas chromatography, problems of
purifying ^{14}C ethylene and assaying radioactive end products are
avoided. On the other hand, the fababean system, because of its high
EM activity, may be an atypical system, and my results may not apply
to other plant systems.

Beyer (7, 8, 9, 10) was the first to provide critical evidence

Y. Fuchs and E. Chalutz (eds.) Ethylene: Biochemical, Physiological and Applied Aspects.
ISBN 90-247-2984-X. Printed in The Netherlands
©1984, Martinus Nijhoff/Dr W. Junk Publishers, The Hague.

that plants oxidized ethylene. He raised the possibility that
ethylene oxidation was a part of, or a reflection of, ethylene
action. He demonstrated that changes in ethylene metabolism were
correlated with abscission, ripening and senescence and that Ag(I)
ions reduced both ethylene action and its conversion to CO_2.

Table 1. Ethylene oxidation by various parts of fababean seedlings.
Tissues from 6-day-old seedlings were used in this experiment and the
initial concentration of ethylene was 2.5 µl/L.

Seedling portion	EM, nl ethylene/g h
Epicotyl tip, 1 cm	4 ± 1
Remaining epicotyl tissue	4 ± 1
Cotyledons	4 ± 1
Upper root, 4 cm	7 ± 1
Root tip, 4 cm	$-3 \pm 5*$

*Negative value = ethylene production.

In M. paraffinicum, the function of ethylene oxidation is to
provide carbon for the growth of this soil inhabiting microorganism.
Many strains have been described that oxidize ethylene and other
alkenes and alkanes (16). M. paraffinicum may play a role in uptake
of hydrocarbon gases by soils (2).

The role for EM in animal systems (22, 23) is unknown.
Nevertheless, a number of workers (23) have reported that rats exhale
ethylene oxide when exposed to ethylene and that
diethyldithiocarbamate (DIECA) inhibited this reaction. DIECA is also
an inhibitor of EM in fababean and M. paraffinicum (unpublished
results). The action of DIECA may be due to the C=S component of the
molecule or to the release of CS_2 during decomposition.

The function of EM in plants is also unknown. Possible
explanations for its presence include: a method synthesizing ethylene
oxide, a regulator of internal ethylene levels, an enzyme system used
in forming hydrocarbon oxides (12), and a consequence of ethylene
interacting with its site of action. The purpose of the experiments
described here was to test the last hypothesis.

Three approaches were used to test the idea that EM was

associated with ethylene action. Experimental details are provided in
the methods section. The first experiment examined the possibility
that silver might block both ethylene action and EM activity. In many
systems, Ag ions (4) block ethylene action. Experiments were
conducted to determine whether fababean seedlings treated with Ag
would exhibit reduced sensitivity to ethylene and reduced levels of EM
activity. A second experiment consisted of comparing the
physiological activity of ethylene analogs with their relative rates
of oxidation. For example, would a relatively inactive ethylene
analog show reduced activity as an EM substrate? A third experiment
utilized CS_2 as a suicide inactivator of EM. Could CS_2
simultaneously block ethylene action and oxidation?

Data in Table 2 show that when 0.6 mM $AgNO_3$ was sprayed on
fababean epicotyls, it had no effect on their EM activity. This
treatment caused a slight inhibition of ethylene action when ethylene
as applied at a non-saturating dose of 0.5 µl/L. When the
concentration of ethylene was increased to 1.0 µl/L the ability of
silver to block ethylene action was lost and as before, no effect on
ethylene oxidation was observed.

Table 2. Silver partially blocks ethylene action but not EM activity.

| Treatment | 0.5 µl/L ethylene | | 1.0 µl/L ethylene | |
	Epicotyl length, mm	EM activity	Epicotyl length, mm	EM activity
Initial	47 C		39 C	
Control	75A	-8 + 0	63 A	-12 + 6
$AgNO_3$	76A	-32 + 18	60 B	-15 + 4
Ethylene	59 B	8 + 14	41 B	36 + 29
$AgNO_3$ + Ethylene	67 AB	18 + 8	42 B	34 + 14

The second experiment tested the relative effects of ethylene
analogs on their oxidation rates, seedling growth, and EM activity of
epicotyls after hydrocarbon treatment. The half-maximal physiological
activity (as µl/L of the hydrocarbon in the gas phase) of ethylene =
0.1; acetylene = 280; vinyl fluoride = 430; vinyl bromide = 1600; and

1,3-butadiene = 500,000 (1). In this experiment we looked for a correlation between the rate of analog oxidation, its effect on the growth of the epicotyl, and any subsequent effect it might have on epicotyl EM activity.

Table 3. Correlation between oxidation and action of ethylene analogs applied to fababeans at a concentration of 10 µl/L.

Hydrocarbon gas	Hydrocarbon uptake µl/g h	Epicotyl length, mm after 24 h	Epicotyls, EM activity, nl ethylene/g h
Initial, 0 h		17 B	
Air control		27 A	1 ± 2
Etnylene	6.2	18 B	27 ± 4
Acetylene	4.2	24 A	6 ± 2
Vinyl fluoride	3.9	28 A	5 ± 1
Vinyl bromide	36.0	26 A	5 ± 1
1,3-butadiene	22.0	23 A	6 ± 2

Table 3 demonstrates no correlation between the relative rates of hydrocarbon oxidation and their effect on epicotyl elongation. Some of the least active growth inhibiting analogs, such as vinyl bromide, were more rapidly oxidized than ethylene. Dodds and Hall (20) have also reported that propylene was more rapidly oxidized than ethylene. The subsequent EM activity of epicotyls was highest in seedlings treated with ethylene when compared to the other analogs tested. This increase in EM activity by ethylene, may represent either, an increased activity of existing EM, or the synthesis of new protein. Vinyl bromide, the most active substrate for EM, was not the most active inducer for EM. This suggests that the substrates were not acting as allosteric effectors but rather that ethylene was acting normonally in inducing the enzyme capable of its own oxidation.

A third way of testing the role of EM in ethylene is to inactivate the enzyme with a suicide substrate or inactivator. As Beyer originally pointed out, CS_2 and COS inhibit etnylene oxidation in plant tissues (5, 8).

While a great deal has been written about the action of CS_2 in microbiol and animal systems, little is known about the effect of this gas in plants. CS_2 inhibits nitrification by _Nitrosomas_ (25, 29). and denatures cytochrome P-450 (27). It inhibits endoplasmic reticulum calcium pumps (26), binds to amino acids and copper to form dithiocarbamates, and inhibits cytochrome oxidase, monoamine oxidase, and alkaline phosphatase (11). Recently, Taylor (30) reported that CS_2 had no effect on photosynthesis and growth, while COS did. I found that CS_2 had no effect on photosynthesis or respiration of fababeans and no effect on endogenous or auxin induced ethylene production by _Phaseolus vulgaris_ L 'Red Kidney' seedlings. In addition, no difference in the rate of CS_2 uptake by viable or heat killed fababean cotyledons was observed. One mM CS_2 in the liquid phase also had no effect on ascorbic acid oxidase, glactose oxidase and uric acid oxidase, while the known copper chelator DIECA, totally inhibited these copper containing enzymes. These observations support the contention that CS_2, because of its structural similarity to ethylene, may act as a suicide inactivator of EM. Ethylene itself has been shown to be a suicide inactivator of cytochrome P-450 in animal systems (28). The effect of CS_2, COS, and CO on ethylene oxidation by fababean seedlings is shown in Table 4.

Inhibition of EM by a short, low level exposure to CS_2 was complete and irreversible. Table 4 shows that the relative order of activity of these inhibitors was S=C=S > S=C=O > C=O. Similar results were obtained in a study of the action of these gases on ethylene oxidation by _Mycobacterium paraffinicum_. Of the three, CO is considered an ethylene analog. The observation that it inhibits EM is further evidence that EM is not involved in ethylene action.

CO_2 had no effect on EM (data not shown). Since CO_2 often inhibits ethylene action, the anticipated result would have been the observation the CO_2 also inhibited ethylene oxidation.

The action of CS_2 as a suicide inactivator of EM makes it a useful tool in illucidating the role of EM in ethylene action on epicotyl elongation. In the experiments shown below, fababean

80

Table 4. The relative activity of CS_2, COS, and CO as suicide inactivators of EM in fababean seedlings.

CS$_2$ and COS were applied as a 10 min pretreatment. CO was present during the course of the experiment.

ml/L Gas	% Inhibition of ethylene uptake		
	CS_2	COS	CO
0	0	0	0
1	82	31	--
10	95	31	--
100	100	48	--
1,000	99	71	--
5,000	--	--	19
10,000	--	--	50
50,000	--	--	99

epicotyls were first treated with an inhibitory concentration of CS_2, and then exposed to ethylene. In this way, it should be possible to see whether or not ethylene retained its growth inhibiting effects in the presence of inactivated EM.

Table 5. CS_2, a suicide inactivator of EM can block ethylene oxidation without inhibiting ethylene action.

Ethylene treatment, μl/L	Epicotyl length, mm after 3 days growth		Total ethylene consumed, μl ethylene/20 seedlings	
	Control	CS_2	Control	CS_2
Initial	11E			
Air	66 A	66 A	-2	-3
0.1	46 B	40 BC	2	-3
1.0	30 CD	20 DE	52	-7
10.0	16 E	18 E	466	4

Two conclusions can be drawn from Table 5. First EM can be inactivated without affecting ethylene action. Secondly, the rate of ethylene oxidation is a linear function of ethylene concentration.

This linear relationship between the rate of oxidation and ethylene concentration was found to extend to concentrations as high as 1,000 μl/L for both the microbial and plant enzyme systems (data not shown). Had some relationship existed between oxidation and action, one would have anticipated a leveling off of oxidation at the physiologically saturating level of ethylene (10 l/L).

Subsequent experiments were performed in order to better understand the ethylene induced increase in EM activity.

Results of an experiment designed to measure the time required for the induction of EM is shown below. Fababeans were treated with ethylene for various periods of time followed by an assay of epicotyl EM activity.

Table 6. Induction of EM in Fababean epicotyls as a function of the time that they were treated with ethylene.

100 μl/L ethylene, hours	EM activity, nl ethylene/g h
0	8 ± 1
2	12 ± 1
12	18 ± 1
17	23 ± 2
24	25 ± 2

As shown in Table 6, an increase in EM activity was observed after 2 hours of ethylene treatment. Additional increases in EM activity occurred following increased exposure to ethylene.

One approach to understanding the cause of the increase in EM (protein synthesis as opposed to enzyme activation) is to treat fababeans with ethylene oxide, the end product of the oxidation reaction. As shown in Table 7, pretreating fababean seedlings with ethylene oxide did not inhibit the induction of EM by ethylene nor the subsequent activity of EM itself.

Table 7. The effect of ethylene oxide, the reaction product of EM on the induction of EM.
    Fababeans were pretreated with the gases shown for 24 h before assaying the epicotyls for endogenous EM activity.

Treatment, 10 μl/L	EM activity, nl ethylene/g h
Air control	3 ± 1
Ethylene	22 ± 1
Ethylene oxide	1 ± 1
Ethylene oxide + ethylene	25 ± 2

One of the features used to characterize ethylene action is its dose response curve. Typically, 0.1 μl/L is the concentration with a half maximal effect and 10 μl/L the saturating dose. As shown in Table 8, a maximum induction of EM occurred with 10 μl/L and 1 μl/L had more than a half-maximal effect. In this experiment the epicotyls of fababeans were assayed for EM activity 24 hours after the seedlings were exposed to ethylene for 24 hours.

Table 8. Dose response curve for the induction of EM in fababean epicotyls.

μl/L Ethylene for 24 hours	EM activity, nl ethylene/g h
0	9 ± 2
1	25 ± 2
10	32 ± 2
100	30 ± 2

A preliminary way of assessing the role of protein and RNA synthesis in the increase in EM activity is to measure the effect of cycloheximide, an inhibitor of protein synthesis, and actinomycin D, and inhibitor of RNA synthesis, on the ability of ethylene to increase EM activity. Because inhibitor studies are not conclusive, the data shown in Table 9 should be considered as only suggestive evidence that protein, but not RNA synthesis is needed for the increase in EM activity. Additional experiments with labeled intermediates are needed to test this interpretation.

An interesting feature of ethylene action is its ability to control protein synthesis either pre- or posttranscriptionally. Examples of ethylene induced mRNA synthesis followed by protein synthesis include abscission (cellulase), and ripening (polygalacturonase, cellulase) (1). However, in other systems, which can be influenced by ethylene, such as seed germination (19), stress, wounding (1) or senescence (31), the increase in enzyme activity can take place even through actinomycin D is used to block RNA synthesis. The study of ethylene action in activating latent, stored, masked, or performed messenger poses a challenge equal to its effects on regulating the genome.

Table 9. Is the increase in EM activity due to protein and RNA synthesis?

	EM activity, nl ethylene/g h	
Treatment	36 μM Cycloheximide	40 μM Actinomycin D
Air control	3 ± 1	3 ± 1
Inhibitor	2 ± 2	6 ± 1
Ethylene	16 ± 6	18 ± 1
Ethylene + inhibitor	7 ± 3	21 ± 4

In conclusion, the work presented here suggests that while the function of ethylene oxidation is not fully understood, evidence exists that it does not play a rule in either ethylene action, or reflects the action of ethylene in fababeans.

MATERIALS AND METHODS

Fababean (Vicia faba L. 'Diana', a gift of Northern Sales Co. Ltd., 200 Portage Ave. Winnipeg, Canada R3C 3X2) were washed 10 min with 10% Clorox (NaOCl), rinsed with running tap water for 10 min, and then coated with Captan. Seeds were then placed on a 5-cm thick layer of water saturated vermiculite, covered with dry vermiculite treated with Captan (0.5 g/1.5 dm^3) and grown in the dark at 22-28 °C for 4 days. In the experiments shown in Tables 1 and 3, the seeds were grown in 11 x 8 cm plastic containers and treated as undisturbed growing seedlings. In the other experiments, 4-day-old seedlings were removed from the vermiculite, rinsed with deionized water, placed horizontally in a gas tight container on either moist paper or 9 cm Petri plates containing 6 ml of various solutions. Hydrocarbon gas levels were monitored by gas chromatography. EM activity was followed by measuring the rate of ethylene uptake (a - sign connotates production) by 1 g of tissue in a 25 ml flask sealed with a silicon rubber stopper and incubated at room temperature. Ethane was injected in the flask as an internal standard. The + values shown are standard deviations and the letter following data are the mean separation by Duncan's multiple range test, 5% level.

REFERENCES

1. Abeles, FB. 1973. Ethylene in plant biology. Acad. Press, NY.
2. Abeles, FB. 1982. Ethylene as an air pollutant. Agr. and For. Bull. Univ. Alberta 5:4-12.
3. Beyer EM. 1975. ^{14}C-Ethylene incorporation and metabolism in pea seedlings. Nature 255:144-147.
4. Beyer EM. 1976. A potent inhibitor of ethylene action in plants. Plant Physiol. 58:268-271.
5. Beyer EM. 1977. $^{14}C_2H_4$: Its incorporation and oxidation to $^{14}CO_2$ by cut carnations. Plant Physiol. 60:203-206.
6. Beyer EM. 1979. Effect of silver ion, carbon dioxide, and oxygen on ethylene action and metabolism. Plant Physiol. 63:169-173.
7. Beyer EM. 1979. [^{14}C] ethylene metabolism during leaf abscission in cotton. Plant Physiol. 64:971-974.
8. Beyer EM. 1980. Recent advances in ethylene metabolism. In B Jeffcoat, ed., Aspects and prospects of plant growth regulators. Monograph 6 Brit. Plant Growth Reg. Groups pp. 27-38.
9. Beyer EM, Blomstrom DC. 1979. Ethylene metabolism and its possible physiological role in plants. In F Skoog, ed., Plant growth substances. Springer-Verlag Berlin pp. 208-218.
10. Beyer EM, Sundin O. 1978. $^{14}C_2H_4$ metabolism in morning glory flowers. Plant Physiol. 61:896-899.
11. Brieger H. 1967. Carbon disulfide in the living organism. In: H. Brieger and J. Teisinger, eds., Toxicology of carbon disulfide. Excerpta Medica Foundation pp. 27-31.
12. Croteau R, Kolattukudy PE. 1974. Direct evidence for the involvement of epoxide intermediates in the biosynthesis of the C-18 family of cutin acids. Arch. Biochem. Biophys. 162:471-480.
13. Davis JB, Chase HH, and Raymond RL. 1956. Mycobacterium paraffinicum n. sp. a bacterium isolated from soil. Appl. Microbiol. 4:310-315.
14. De Bont JAM. 1975. Oxidation of ethylene by bacteria. Ann. Appl. Biol. 81:119-121.
15. De Bont JAM. 1976. Oxidation of ethylene by soil bacteria. Antonie Van Leeuwenhoek 42:59-71.

16. De Bont JAM, Albers RAJM. 1976. Microbiol metabolism of ethylene. Antonie Van Leeuwenhoek 42:73-80.

17. De Bont JAM, Atwood MM, Primrose SB, Harder W. 1979. Epoxidation of short chain alkenes in Mycobacterium E20: The involvement of a specific monooxygenase. FEMS Microbiol Letts. 6:183-188.

18. De Bont JAM, Harder W. 1978. Metabolism of ethylene by Mycobacterium E20. FEMS Microbiol Letts. 3:89-93.

19. Dommes J, Van de Walle C. 1983. Newly synthesized mRNA is translated during the initial imbibition phase of germinating maize embryo. Plant Physiol. 73:484-487.

20. Dodds JH, Hall MA. 1982. Metabolism of ethylene by plants. Int. Rev. Cytol. 76:299-325.

21. Dodds JH, Musa SK, Jerie PH, Hall MA. 1979. Metabolism of etnylene to ethylene oxide by cell-free preparations from Vicia faba L. Plant Sci. Letts. 17:109-114.

22. Enrenberg L, Osterman-Golkar S, Segerback D, Svensson K, Calleman CG. 1977. Evaluation of genetic risks of alkylating agents. III Alkylation of hemoglobin after metabolic conversion of ethene to ethylene oxide in vivo. Mutation Res. 45:175-184.

23. Filser JG, Bolt HM. 1983. Exhalation of ethylene oxide by rats on exposure to ethylene. Mutation Res. 120:57-60.

24. Jerie PH, Hall MA. 1978. The identification of ethylene oxide as a major metabolite of ethylene in Vicia faba L. Proc. R. Soc. Lond. B. 200:87-94.

25. Kudeyarov VN, Jenkinson DS. 1976. The effects of biocidal treatments on metabolism in soil. VI. Fumigation with carbon disulfide. Soil Biol. Biochem. 8:375-378.

26. Moore, L. 1982. Carbon disulfide hepatoxicity and inhibition of liver microsome calcium pumps. Biochem. Pharmacol. 31:1465-1467.

27. Obrebska MJ, Kentish R, Parke DV. 1980. The effects of carbon disulfide on rat liver microsomal mixed-function oxidases, in vivo and in vitro. Biochem. J. 188:107-112.

28. Ortiz De Montellano PR, Beilan HA, Kunze KL, Mico BA. 1981. Destruction of cytochrome P-450 by ethylene. J. Biol. Chem. 256:4395-4399.

29. Powlson DS, Jenkinson DS. 1971. Inhibition of nitrification in soil by carbon disulfide from rubber bungs. Soil Biol. Biochem. 3:267-269.

30. Taylor GE. 1983. The significance of developing energy technologies of coal conversion to plant productivity. HortSci. 18:684-689.

31. Thomas H, Stoddart JL. 1980. Leaf senescence. Ann. Rev. Plant Physiol. 31:83-111.

32. Wiegant WM, De Bont JAM. 1980. A new route for ethylene glycol metabolism in Mycobacterium E 44 J. Gen. Microbiol. 120:325-331.

SUPERINDUCTION OF ACC SYNTHASE IN TOMATO PERICARP BY LITHIUM IONS *

Thomas Boller

Botanisches Institut der Universität Basel,
Schönbeinstrasse 6, CH-4056 Basel, Switzerland

INTRODUCTION

In attempts to purify ACC synthase, it is important to take a tissue with high specific activity of the enzyme as a starting material. The pericarp of ripening tomato fruit, wounded by cutting, has yielded comparatively high specific activities of the enzyme (1) and, therefore, has been the material c choice for biochemical work on ACC synthase (2).

However, in absolute terms, the activity of ACC synthase is quite low ever in wounded tomato fruit. Furthermore, it proved to be difficult to obtain the enzyme in good yield from larger batches of wounded tomato pericarp (2). Here I report on a possibility to induce ACC synthase in tomato fruit further.

MATERIAL AND METHODS

Discs from tomato pericarp (diameter 11 mm, 3 mm thick) were incubated on filter paper wetted with the test solutions. The tissue was powdered in liqui nitrogen and extracted with 100 mM phosphate buffer, pH 8, containing 2 mM dithiothreitol and 2 μM pyridoxal phosphate. ACC synthase was measured in the tissue extracts after desalting on Sephadex G50 columns as described (1).

RESULTS

Effect of various chemicals on the induction of ACC synthase by wounding

The excision of discs from tomato tissue strongly induces ACC synthase (1) Therefore, discs incubated on water had a high level of ACC synthase. This level could not be increased further by chemical treatments known to induce ACC synthase in other tissues, e.g. $CdCl_2$ (3), AVG (4), or $AgNO_3$ with auxin ε kinetin (5). However, incubation on 50 – 100 mM lithium salts (LiCl, LiBr, $LiNO_3$) caused a further 4 – 5fold induction of ACC synthase above the level ir duced by wounding. Other alkali salts (NaCl, KCl, RbCl) were ineffective.

---

* I thank R. Schweizer for technical help. This work was supported by a grant from the E.Schnurr-Guggenheim-Stiftung, Basel.

Y. Fuchs and E. Chalutz (eds.) Ethylene: Biochemical, Physiological and Applied Aspects.
ISBN 90-247-2984-X. Printed in The Netherlands
©1984, Martinus Nijhoff/Dr W. Junk Publishers, The Hague.

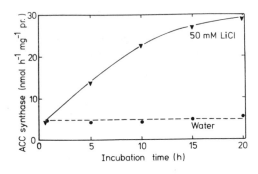

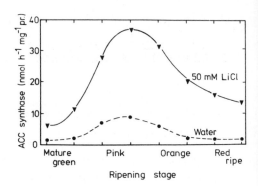

FIGURE 1. Time course of superinduction of ACC synthase by lithium chloride.

FIGURE 2. Dependence of ACC synthase induction by wounding or LiCl on the ripening stage of the tomato fruit.

Characterization of the superinduction of ACC synthase by lithium chloride

In pericarp discs incubated on water, the level of ACC synthase increased dramatically in the first hour (not shown) and then remained approximately constant (Fig. 1). In discs incubated on 50 mM LiCl, the activity of ACC synthase continued to increase steadily for 20 h (Fig. 1).

The induction of ACC synthase by cutting depends on the ripening stage of the fruit (1). The superinduction by LiCl similarly depended on the ripening stage (Fig. 2). Compared to the level attained by wounding, LiCl further increased ACC synthase 5fold in the earlier and 10fold in the later ripening stages (Fig. 2).

DISCUSSION

Treatment of tomato pericarp with LiCl for 15 - 20 h causes a superinduction of ACC synthase above the already high level induced by wounding. The treatment is particularly effective in ripe tomato fruit.

The LiCl treatment can be scaled up by filling hollowed fruit halves with a 50 mM LiCl solution. Several kg of tissue were treated in this way and yielded an extract with a specific ACC synthase activity of 20 - 30 nmol h^{-1} mg^{-1} protein.

In conclusion, tomato pericarp treated with LiCl is an excellent source for ACC synthase in attempts to characterize and purify the enzyme.

REFERENCES

1. Kende H, Boller T, 1981. Planta 151, 476 - 481.
2. Acaster MA, Kende H, 1983. Plant Physiol. 72, 139 - 145.
3. Fuhrer J, 1982. Plant Physiol. 70, 162 - 167.
4. Yoshii H, Imaseki H, 1982. Plant Cell Physiol. 23, 639 - 649.
5. Aharoni N, Anderson JD, Lieberman M, 1979. Plant Physiol. 64, 805 - 809.

1-AMINOCYCLOPROPANE-1-CARBOXYLIC ACID (ACC)-DEPENDENT ETHYLENE SYNTHESIS IN VACUOLES OF PEA AND BEAN LEAVES*

M. GUY, H. KENDE

MSU-DOE Plant Research Laboratory, Michigan State University, East Lansing, MI 48824, USA

Protoplasts formed ACC and evolved ethylene during incubation. ACC and ethylene synthesis were promoted by light and inhibited by DCMU, CCCP and AVG. When protoplasts were incubated on L-[U-^{14}C]methionine, the specific radioactivity of ACC increased much more rapidly than that of ethylene (1).

Vacuoles isolated from protoplasts at different time points of incubation contained between 80% and 85% of the total cellular ACC. The ACC level in vacuoles mirrored the increase in the ACC level of protoplasts. When vacuoles were isolated from freshly prepared protoplasts and incubated for 6 h, the ACC level of the vacuoles increased very little, if significantly at all, while that of the corresponding protoplasts rose almost fivefold. Isolated vacuoles formed ethylene. The amount of ethylene produced by isolated vacuoles was 75% to 86% of that formed by protoplasts (2).

Vacuoles isolated from AVG-treated protoplasts were used to study the kinetics and stereospecificity of ACC-dependent ethylene biosynthesis. Ethylene synthesis was saturated at 1 mM ACC, the apparent Km for the conversion of ACC to ethylene was 61 μM and the Vmax 32 pmol/h/10^6 vacuoles (2). Hoffman et al. (3) reported that one of the the four stereoisomers of 1-amino-2-ethylcyclopropane-1-carboxylic acid (AEC), (1R,2S)-AEC or (1) allocoronamic acid, was preferentially converted to 1-butene by apple

---

*Supported by the U.S. Department of Energy under contract No. DE-AC02-76ER01338.

Y. Fuchs and E. Chalutz (eds.) Ethylene: Biochemical, Physiological and Applied Aspects.
ISBN 90-247-2984-X. Printed in The Netherlands
©1984, Martinus Nijhoff/Dr W. Junk Publishers, The Hague.

and mungbean hypocotyl tissue. They suggested that 1-butene was formed by the same enzyme as ethylene. Pea vacuoles incubated on (+)-allocoronamic acid formed 1-butene, those incubated on (+)-coronamic acid did not. ACC inhibited the formation of 1-butene from (+)-allocoronamic acid, and (+)-allocoronamic acid inhibited the formation of ethylene from ACC (2).

Ethylene formation from endogenous ACC was greatly inhibited in isolated vacuoles by $CoCl_2$, n-propyl-gallate and in an atmosphere of N_2. Ethylene synthesis also ceased when vacuoles were lysed by passage through a syringe and hypodermic needle (2).

Vacuoles prepared from Vicia faba protoplasts produced ethylene much like pea vacuoles did, and more than 75% of the ACC per protoplast was sequestered in the vacuole.

CONCLUSIONS
a. ACC is synthesized in the cytoplasm of pea and Vicia faba protoplasts, and a major portion of it is subsequently sequestered in the vacuole.
b. Isolated vacuoles are not capable of forming ACC but convert ACC to ethylene. Over 80% of the ethylene produced by protoplasts could be accounted for as originating from the vacuole.
c. ACC-dependent ethylene synthesis in isolated vacuoles and in intact plant tissue share a number of important similarities: The Km, saturation above 1 mM ACC, stereospecificity, inhibition by $CoCl_2$, n-propyl-gallate and under N_2.
d. We do not conclude that the vacuole is the sole site of ethylene synthesis. Indeed, ethylene synthesis may also occur on a cytoplasmic membrane or the plasmalemma as has been suggested by others.

REFERENCES
1. Guy, M., Kende, H. Planta (in press).
2. Guy, M., Kende, H. Planta (in press).
3 Hoffman, N.E., Yang, S.F., Ichihara, A., Sakamura, S. 1982. Plant Physiol. 70:195-199.

ACC CONVERSION TO ETHYLENE BY MITOCHONDRIA FROM ETIOLATED PEA (PISUM SATIVUM) SEEDLINGS

CHANA VINKLER AND AKIVA APELBAUM
DEPT. FRUIT AND VEGETABLE STORAGE, ARO, THE VOLCANI CENTER, BET DAGAN, ISRAEL

The ability of plant mitochondria to convert 1-aminocyclopropane-1-carboxylic acid (ACC) to ethylene was measured in mitochondria from 7 day-old etiolated pea seedlings. Previous attempts to measure ethylene formation by mitochondria (1,2) were done in the absence of ACC.

Sonicated submitochondrial particles (SMP) are inside-out vesicles of the inner mitochondrial membrane. When SMP were prepared from isolated and purified mitochondrial particles 2-4 fold increase in activity was obtained (Figs. 1 and 2). This activity was almost completely recovered from the pellet. Thus, ethylene formation was associated with the inner face of the inner mitochondrial membrane and did not represent soluble activity in the mitochondrial matrix (Fig. 1). Our studies indicate that the rate limiting step in the activity of ACC conversion to ethylene is the permeation of ACC to the inner side of the mitochondria and upon sonication this limitation is eliminated, due to the exposure of the inner face of the inner mitochondrial membrane to the medium (3).

Further support to this idea comes from treatment of the intact mitochondria with specific ionophores or with detergents which make the mitochondrial membrane more permeable to ACC. Fig. 2 shows that an 8-fold increase in activity of ACC conversion to ethylene is obtained with cholate added to mitochondria, whereas no effect is observed when cholate is added to SMP. At high cholate concentrations the membranes are completely solubilized and inactivation occurs, both in intact mitochondria and in SMP. Similar effects were observed by treatment with ionophores like valinomycin or gramicidine and with the detergent deoxycholate.

The conversion reaction was inactivated by heat (10 min, $100^{\circ}C$), was Mn^{+2} dependent, has an optimal pH value at pH 8.0, is highly sensitive

Y. Fuchs and E. Chalutz (eds.) Ethylene: Biochemical, Physiological and Applied Aspects.
ISBN 90-247-2984-X. Printed in The Netherlands

Discs sliced from mesocarp of winter squash fruit produced ethylene at an increasing rate after a lag period of 3 h upon incubation at 25°C. ACC synthase activity increased at a rapid rate after a lag period of 2 h, reaching a peak 6 to 14 h after the beginning of incubation and then declining sharply (1). The time when ACC synthase activity reached a peak and the extent of its maximum activity (50 to 156 nmolg^{-1}h^{-1}) varied with the storage period and condition after harvest and with individual fruits. The rise in ACC synthase activity was followed by a rapid increase in ACC formation and ethylene production (1). EFE activity also increased at a rapid rate. Induction of ACC synthase by wounding was strongly suppressed' by the application of inhibitors of protein and nucleic acid synthesis. Arrhenius plots of the rate of ethylene production by the wounded tissue showed discontinuity at 9.9°C, suggesting the involvement of the membrane structure in the formation of ethylene. The induction of ACC synthase by wounding was suppressed by ethylene. On the other hand, the induction of EFE was not suppressed but stimulated by ethylene (Table 1).

Table 1. Effect of ethylene on induction of ACC synthase, ACC formation and EFE in the wounded tissue of mesocarp of winter squash fruit.

Expt.	Treatment	ACC synthase (nmolg^{-1}h^{-1})	ACC content (nmolg^{-1})	C_2H_4 production (nmolg^{-1}h^{-1})	EFE (nmolg^{-1}h^{-1})
1	Control	33.0	41.0	3.5	—
	C_2H_4 (114 µl/1)	8.4	18.8	3.0	—
2	Control	—	—	—	0.73
	C_2H_4 (104 µl/1)	—	—	—	0.93

Discs were incubated for 7 h (Expt. 1) and 5 h (Expt. 2), respectively.

Conclusion: Wounding markedly induced ACC synthase and EFE in mesocarp of winter squash fruit, leading to a marked formation of ACC and ethylene. Ethylene suppressed the induction of ACC synthase but enhanced the induction of EFE.

REFERENCE

1. Hyodo H, Tanaka K, Watanabe K. 1983. Wound-induced ethylene production and 1-aminocyclopropane-1-carboxylic acid synthase in mesocarp tissue of winter squash fruit. Plant & Cell Physiol. 24:963-969.

CHARACTERIZATION OF THE CARBOHYDRATES-STIMULATED ETHYLENE PRODUCTION IN TOBACCO LEAF DISCS

SONIA PHILOSOPH-HADAS, SHIMON MEIR AND NEHEMIA AHARONI
Agricultural Research Organization
The Volcani Center
Bet Dagan, Israel

Unlike fruits (3), intact leaves were found to produce very small amounts of ethylene (1). Besides by stress effects and addition of growth regulators (3), these low ethylene production rates could be increased also by employing several sugars (2,5). We have found that twelve naturally occurring carbohydrates could remarkably stimulate ethylene production rates in tobacco leaves. The present investigation demonstrates: (a) Characterization of the sugar-stimulated ethylene production system in tobacco leaf discs; and (b) Sites of sugar-stimulation in the ethylene biosynthesis pathway.

Methods. Experiments were performed with samples of 8 tobacco leaf discs (Nicotiana tabacum L. cv. Xanthi) incubated in 50-ml flasks containing 2 ml of 50 mM Na-phosphate buffer, pH 6.1, in a CO_2 and ethylene-free atmosphere. Ethylene absorbed by $Hg(ClO_4)_2$ was released with saturated LiCl (6) and measured by gas chromatography. Where indicated, 50 mM sugars, 0.1 mM IAA, 0.1 mM ACC (1-amino-cyclopropane-1-carboxylic acid), 0.1 mM AVG (aminoethoxyvinylglycine(2-amino-4-(2'-aminoethoxy)-trans-3-butenoic acid), 0.5 mM Co^{2+} or 1 μCi L-(3,4-^{14}C)methionine were included in the incubation medium. ACC extraction and assay were performed according to the method of Lizada and Yang (4). For the radioactive measurements, the absorbed $^{14}C_2H_4$ was liberated by LiCl (6), reabsorbed by $Hg(C_2H_3O_2)_2$ solution and then directly counted by liquid scintillation.

Results: Sucrose or glucose applied at a concentration range of 1 to 100 mM to IAA-treated or untreated tobacco leaf discs, stimulated ethylene production for several days. Among the 14 carbohydrates tested, mannitol had no effect while galactose and sucrose were the most active in stimulating ethylene production by 10-fold after an apparent lag period of 12 h from excision. However, the pattern of stimulation during incubation in galactose-treated discs differed from that of glucose-treated ones.

Y. Fuchs and E. Chalutz (eds.) Ethylene: Biochemical, Physiological and Applied Aspects.
ISBN 90-247-2984-X. Printed in The Netherlands
©1984, Martinus Nijhoff/Dr W. Junk Publishers, The Hague.

98

enhanced their hydrolysis. Reapplication of each of the 3 IAA conjugates (at 10 µM) in the presence of sucrose (50 mM), resulted in a marked stimulation of ethylene production and decarboxylation by the leaf discs, when the esteric IAA conjugate was the most active, giving a four-fold ethylene enhancement as compared with the free IAA applied (0.1 mM). The metabolism of the 2 IAA conjugates was identical to that of free IAA. In pulse experiments, where 0.1 mM IAA was applied only for 4 h, in the presence of sucrose, increased ethylene production continued for several days to a similar extent as in the experiments with a continuous supply of IAA.

Tracer experiments with (^{14}C)IAAla (10 µM) revealed that this synthetic IAA conjugate can markedly induce ethylene production in tobacco leaf discs when applied with sucrose, and its metabolism resembles that of free IAA. On the other hand, when applied without sucrose, a large amount of (^{14}C)IAAla was retained in the leaf tissue and did undergo neither oxidation nor conjugation.

Conclusions: (a) Extracts of tobacco leaf discs contained 84% of their endogenous IAA derivatives as esteric IAA. (b) Sucrose stimulates hydrolysis of both the esteric and the peptidic IAA conjugates, formed in the tissue after application of free IAA. (c) All the 3 IAA conjugates can induce ethylene production in sucrose-treated tobacco leaf discs, when the esteric conjugate is the most active. (d) Both the endogenous IAA conjugates and synthetic IAAla exert their biological activity by releasing free IAA in the tissue, as expressed by their decarboxylation and their turnover pattern. This activity is pronounced mainly with sucrose. (e) Sucrose can elicit a continuous IAA effect for 48 h, as expressed by increased ethylene production, even when IAA was removed from the medium after 4 h. (f) The results suggest that sucrose increases ethylene production in tobacco leaf discs by stimulating the hydrolysis of either endogenous or exogenously supplied IAA conjugates, thereby causing a slow release of free IAA.

References:

1. Aharoni N and Yang SF. 1983. Plant Physiol. 73:598-603.
2. Epstein E and Cohen JD. 1981. J. of Chromatography 209:413-420.
3. Feung CH, Hamilton RH, Mumma RO. 1977. Plant Physiol. 59:91-93.
4. Hangarter RP, Peterson MD, Good NE. 1980. Plant Physiol. 65:761-767.
5. Philosoph-Hadas S, Meir S, Aharoni N. 1983. Plant Physiol. Suppl. 72:121.

THE EFFECT OF CYCLOALKENES ON ETHYLENE BINDING

A.R.SMITH, C.J.HOWARTH, I.O. SANDERS & M.A.HALL

Department of Botany & Microbiology, University College of Wales,
Aberystwyth, Dyfed SY23 3DA, U.K.

INTRODUCTION

The effects of cycloalkenes on ethylene binding have been
investigated using cell-free preparations of *Phaseolus vulgaris*
L. cotyledons and leaf disks of *Nicotiana tabacum* L.

MATERIALS AND METHODS

Leaves of *N. tabacum* L. cv. Wisconsin 38 were harvested when
20-30cm long and surface sterilised with 5% w/v NaOCl. Leaf
disks 1cm in diameter were cut and placed on moist filter paper
for 1h prior to assay. Plants of *P.vulgaris* L. cv. Canadian
Wonder were grown as described previously (1). Isolation of
the cell-free system, mode of incubation with ethylene and
assays for ethylene binding and metabolism were as described
previously (1,2,3).

RESULTS AND DISCUSSION

Cyclopentene, 2,5-norbornadiene, 1,3-cyclohexadiene and
1,3-cycloheptadiene all inhibited binding of $^{14}C_2H_4$ competitively in the cell-free preparation from *P. vulgaris*. However,
neither norbornene nor norbornane had any effect on $^{14}C_2H_4$
binding. Table 1 shows the relative inhibitor constant (K_i)
for each of the cycloalkenes tested in the *P. vulgaris* cell-free
binding system. When K_i values of these compounds are compared
to the K_D for ethylene binding their relative effectiveness as
inhibitors is seen to be very low. Therefore the potential of
2,5-norbornadiene and other cycloalkenes tested in ethylene
binding studies is limited.

Table 2 shows that tissue incorporation of ^{14}C was detected
in tobacco leaf disks but at very low levels. However more
than five times as much ^{14}C was detected in NaOH traps than

Y. Fuchs and E. Chalutz (eds.) Ethylene: Biochemical, Physiological and Applied Aspects.
ISBN 90-247-2984-X. Printed in The Netherlands
©1984, Martinus Nijhoff/Dr W. Junk Publishers, The Hague.

Table 1. Comparison of inhibitor constants for cycloalkenes in the *P. vulgaris* cell-free ethylene binding system. Samples incubated with $^{14}C_2H_4$ at a concentration range of $0.6.10^{-9}$-$13.0.10^{-9}M$ (liquid phase) in the presence or absence of cycloalkene. K_i values calculated from Scatchard plots.

	Partition Coefficient	K_i (Gas) Relative	K_i (Liquid) Relative
Ethylene	0.11	1	1
Cyclopentene	0.30	10490	28606
2,5-Norbornadiene	0.98	10920	91670
1,3-Cyclohexadiene	1.43	17250	156975
1,3-Cycloheptadiene	1.70	21130	323310

in tissue and 99% of the radioactivity was lost upon acidification indicating the conversion of $^{14}C_2H_4$ to $^{14}CO_2$.

Table 2. ^{14}C-ethylene incorporation/metabolism by tobacco leaf disks. Effect of cycloalkenes. Samples incubated with a constant $^{14}C_2H_4$ concentration of $6.10^{-9}M$ (liquid phase). All values for 1g fresh mass of tissue. Values obtained with heat killed tissue shown in brackets.

	Tissue Incorp. dpm		NaOH Soluble ^{14}C dpm		Metabolism to CO_2 dpm	
^{14}C-C_2H_4 Alone	120	(54)	630	(53)	626	(63)
+ Norbornene						
2142 μll^{-1}gas phase	126	(60)	639	(58)	621	(62)
+2,5-Norbornadiene						
2160 μll^{-1}gas phase	58	(55)	659	(59)	642	(76)
$^{14}C_2H_4$ Alone-*P.vulgaris*						
cell-free preparation	23259		-		-	

Neither 2,5-norbornadiene nor norbornene inhibited $^{14}CO_2$ formation although 2,5-norbornadiene inhibited tissue incorporation. As the formation of $^{14}CO_2$ from $^{14}C_2H_4$ in tobacco leaf disks does not show Michaelis-Menten kinetics it is therefore difficult to assess the physiological significance of this conversion.

REFERENCES

1. Bengochea, T., Dodds, J.H., Evans, D.E., Jerie, P.H., Niepel B., Shaari, A.R. & Hall, M.A. 1980. Planta *148*, 397-406.
2. Bengochea, T., Acaster, M.A., Dodds, J.H., Evans, D.E., Jerie, P.H. & Hall, M.A. 1980. Planta *148*, 407-411.
3. Beyer, E.M. 1975. Nature *255*, 144-147.

ETHYLENE AND FLOWER SENESCENCE

R. NICHOLS
Glasshouse Crops Research Institute, Worthing Road, Littlehampton,
W. Sussex, U.K.

INTRODUCTION

The flower is a complex organ consisting of reproductive and vegetative tissues, each with differing rates of metabolism and senescence,
which respond differently to exogenous ethylene. Furthermore, abscission
layers may occur in some parts of the flower so that entire organs may
abscind without obvious symptoms of the effects of ethylene on the component tissues. Thus flower senescence is a subjective term and includes
a range of phenomena from loss of turgour of individual petals to the
shedding of an entire corolla.

The specific interest in ethylene in relation to flower senescence
started years ago, when it was recognised that gaseous pollutants,
notably from illuminants, caused symptoms on flowers that were commercially unacceptable (1). Once ethylene was identified as one of the
physiologically active components of such gases, extensive studies were
made to describe its effect on a wide range of species and review papers
have been published (2, 3, 4). Particular aspects are dealt with in
general reviews of ethylene in plant metabolism (5, 6, 7). Halevy and
Mayak(8, 9) have reviewed the papers relevant to an understanding of the
factors that contribute to flower senescence and included a comprehensive
account of the part played by ethylene. In this paper the recent
findings concerned with ethylene in flowers are discussed.

Practical aspects of ethylene as a pollutant

There are sources other than those of illuminant gases which are
capable of generating potentially phytotoxic levels of ethylene such as
emissions from internal combustion engines, components of the volatiles
from ripening fruits and products from the breakdown of rubber and
insulating materials (10). Concentrations of ethylene found in the

Y. Fuchs and E. Chalutz (eds.) Ethylene: Biochemical, Physiological and Applied Aspects.
ISBN 90-247-2984-X. Printed in The Netherlands
©1984, Martinus Nijhoff/Dr W. Junk Publishers, The Hague.

atmosphere have been measured in various locations and under a variety of climatic conditions; Abeles cites a number of papers concerned with this topic (10). There are three important considerations which have to be borne in mind: the effect of dosage (time x concentration), temperature and the physiological age of the tissue.

It has been the view that flowers in general are sensitive to ethylene but this may have to be qualified because it is becoming evident that there are substantial differences between species. How far this can be taken is difficult to assess because from the available evidence it is not possible to ensure that flowers at similar stages of development are being compared. Much of the work on the reaction of flowers to known doses of ethylene has been done with cut flowers as selected by market criteria. The findings do not help to clarify the true relative sensitivities of flowers to ethylene unless the stage of development of the inflorescence is taken into account. Flowers, like fruits, are much less responsive to ethylene when they are immature. The behaviour of the carnation flower, much studied because of its commercial importance and apparent sensitivity to ethylene, illustrates the point. If a mature flower, with outer petals reflexed as defined by market standards, is exposed to 0.5 to 1.0 vpm ethylene for 24 h, the outer petals in-roll, all petals lose turgour and then wilt irreversibly. These are the obvious visible symptoms although a sequence of other metabolic events is initiated which leads to accumulation of dry matter in and enlargement of the ovary. In contrast, immature flowers (buds) will tolerate such treatment without much effect on their longevity (11, 12) and at higher doses of ethylene the styles and petals elongate (13). In addition, the ambient temperature when the flowers are exposed to ethylene is also important: at low temperatures, flowers are more tolerant to a given ethylene exposure (11) than at higher temperatures. Season and water status also affect ethylene responses perhaps by influencing the osmotic potential of the tissues (14).

It is clear that the responses of flowers are complex. The main difficulty is to identify a realistic concentration of ethylene that is found in practice and to relate it to the time of exposure of the flowers and to the other considerations outlined earlier. Nonetheless, a description of flower responses is useful for diagnostic purposes, provided the limitations are recognised. Examples of ethylene effects are: dry sepal

of orchid (15); abscission of antirrhinum and calceolaria flowers (16);
reversible, or more importantly, irreversible wilting of carnations (17),
wilting of Kalanchoe (18) and Alstroemeria (23); anthocyanin leakage in
Tradescantia (19); in-rolling of corollas of Ipomoea (20); colour changes
in orchids (21); abscission of florets of Zygocactus (22) and Digitalis
(24).

Ethylene as a hormone in flower senescence

Several authors have referred to ethylene as a hormone or indicated
that it plays a role in plant growth and development (5, 7, 25). It
seems reasonable, therefore, that it has such a role in flower development. The observation that by altering the endogenous levels of ethylene
with exogenous ethylene leads to symptoms analogous to those of accelerated of natural senescence has supported this view. It has been shown
that an ethylene surge is consistently associated with terminal senescence of the carnation flower (17, 26, 27). The similarity between these
physiological changes namely enhanced autocatalytic ethylene production,
accompanied by increased respiration, and those of fruit ripening,
suggested that there is in floral physiology a system similar to that
functioning in some fruits. Some flowers, as with fruits, do not show
this climacteric-type rise in ethylene production, but it may not be
rewarding to take the analogy with fruits too far. However, it is evident that many of the chemical treatments that promote or retard ripening
of climacteric fruits also accelerate or delay senescence of flowers,
particularly for those species which exhibit the ethylene surge at senescence. Abscisic acid (28) or auxins (29, 30) accelerate senescence of
carnation flowers with enhanced ethylene production whereas kinins delay
it (31, 32). Reports concerned with effects of growth substances on
flowers tend to be inconsistent but some of the anomalies can be explained by differences in absorption and translocation of such compounds
into and through the vascular system. In our experiments, accelerated
wilting of carnation flowers with 30 cm stems did not always occur if the
stems were placed in 1.0 mM IAA whereas flowers with 5.0 cm stems wilted
consistently. It was presumed that the IAA was bound or oxidized during
its passage through the longer stem; 2:4D promoted wilting of flowers
with stems of either length.

Metabolism of ethylene

There seems little doubt that ethylene is metabolized in flowers (33,

34). Compounds which block possible sites of ethylene action inhibit the reaction of flowers to ethylene. The evidence from flower studies indicates that the pathway of ethylene biosynthesis is similar to that described by Yang (35):

methionine – S-adenosylmethionine (SAM) – 1-aminocyclopropane-1-carboxylic acid (ACC) – ethylene.

Inhibitors of the pathway such as aminoethoxyvinylglycine (AVG) or aminooxyacetic acid (AOA) inhibit ACC production and hence ethylene synthesis, and treatments with these compounds increase flower longevity (4, 36, 37). Treatments with ACC, like ethylene, shorten flower life (38, 39). Silver ions, which inhibit effects of ethylene (40), either increase flower longevity or counter the deleterious effects of exogenous ethylene; in this respect, silver complexed as the thiosulphate salt (STS) is particularly effective, probably as a result of its greater mobility in the anionic form (41, 42). It appears that silver ions will block the effects of exogenous ethylene even in those species, roses and tulips, which do not usually show a marked response to silver ions in the absence of ethylene (43, 44). In this context, the response by tulip cut flowers is essentially a reversal of the ethylene-induced suppression of stem elongation after a pre-treatment with silver; the mature tulip inflorescence seems relatively insensitive to ethylene. The rose corolla exhibits a rise in ethylene towards the end of senescence (45) although smaller than that reported for carnation. It follows that these flowers absorb, translocate and presumably bind sufficient silver to block the activity of ethylene. In the carnation, silver blocks the surge of autocatalytic ethylene but not the ethylene synthesis associated with normal cell metabolism; ACC also rises as the petals wilt irreversibly and its production is also suppressed by silver (46).

However, the role of ethylene and by inference the effects of silver are complex. Sufficient silver thiosulphate is absorbed by lily bulbs to affect flower quality some weeks after the treatment (47, 48, 49) which implies that the silver is absorbed into the bulb and influences growth and development of the inflorescence. It is clear, however, that immature flowers can be affected by ethylene since storage of tulip bulbs in an ethylene-contaminated atmosphere can lead to a variety of disorders of the flower (50) and in lilies leads to necrosis or abscission of the bud of the emerged flower. The ethylene sensitivity of the immature flower may explain the desirable effects of the silver treatment.

Studies by electron microscopy of carnation flowers treated with STS revealed silver-containing particles in cell walls and plasmodesmata (51). In carnation (52) and tulip stems (Nichols and Atkey, unpublished) silver deposits in middle lamellae or in close proximity to the plasma-lemma of vascular tissue. Clearly this does not preclude activity of silver ions elsewhere in the cell. There is indirect evidence that ethylene may be metabolized near the plasma membrane and the structural integrity of such membranes has been thought to be important for ethylene metabolism (7). Indeed, there is considerable evidence that properties of flower cell membranes change as they age (17, 19, 53, 54). Intact protoplasts from Ipomoea corollas (55) and microsomal membranes (56) have been shown to convert ACC to ethylene, and treatment with ethylene accelerates membrane senescence (57). It seems possible that the effects of apparently disparate groups of chemicals on senescence phenomena can be explained by their influence on membrane structure or function. Free radical scavengers which inhibit conversion of ACC to ethylene (58) also increase flower longevity of some species (59). In this context lipoxygenase activity which increases during senescence and is associated with free radical production may affect membrane integrity. The delay in senescence of carnation flowers and other tissues treated with allopurinol might be explained by suppression of free radical formation (60). It is possible that polyamines which suppress ethylene production in Tradescantia petals (61) and have been reported to delay flower senescence (62) may exert their influence on membrane function.

The role of ethylene in flower senescence

It seems that ethylene functions in flowers as it does in other tissues, namely as part of a growth regulator complex presumably controlling development and function. In certain flowers it also seems to coordinate the sequence of events ending senescence whether this is by abscission or irreversible wilting; shedding or death of the flower is accelerated by ethylene. In Vanda an increase in ethylene production was found in response to pollination or auxin (63) and substances additional to auxin have been suggested to regulate post-pollination phenomena in orchid flowers (64). The stimulation of ethylene from flowers by pollen has been reported for a range of species suggesting that the phenomenon is common in plants. The style appears to be an active site of ethylene synthesis and pollen stimulates ACC and ethylene production between 1 and

4 h after contact with the stigma (65). Pollens from a range of species contain ACC and in carnations sufficient to account for a small surge of ethylene shortly after the pollen and stigma make contact (66). The role of ACC in this context is of particular interest. There is evidence that a senescence-accelerating compound diffuses from the bases of aged carnation gynoecia (67) which will cause in-rolling of detached petals; ACC has similar effects (67, 38). Pollination evokes a comparable petal response in the intact carnation (65). ACC is translocated in some vegetative tissues (35) but this has yet to be demonstrated for floral organs.

We have used the Petunia flower in our more recent work. It has a pattern of senescence which resembles that of the carnation except that the entire corolla will detach from the receptacle at the end of irreversible wilting; in the carnation, petals shrivel but remain attached to the receptacle. In both species pollination will induce accelerated irreversible wilting. Style damage accelerates corolla wilting of Petunia (68). However, it has been shown that pollination or style wounding promotes ethylene production from the gynoecium and rapid wilting of the Petunia corolla; the response to pollen or wounding could be inhibited by STS, implicating involvement of ethylene (69).

We have observed that wounding the stigma of Petunia flowers, attached to the plant, causes irreversible corolla wilting within 2 to 3 days (Nichols and Frost, unpublished data) which is accompanied by a rise in levels of ACC in the gynoecium. Since the stigma is about 2.0 cm distal from the corolla base, a hormone-like signal must pass to the corolla through the receptacle. Pretreatment with AVG (1.0 mM) prevented the wound-induced corolla response suggesting that ACC synthesis was involved. In these experiments with Petunia there was no evidence of swelling of the ovary unless the undamaged stigma was pollinated. In contrast, accelerated corolla wilting of the carnation is generally accompanied by swelling of the ovary and receptacle, regardless of the method (germinating pollen, auxin, ACC, ethylene, propylene) by which it is induced. Thus the stimulus for ovary growth in the Petunia is derived from products of fertilization, presumably as a result of seed induction, and is not the same as that which evokes corolla wilting. In the carnation, corolla wilting and early growth of the ovary can be induced by the same or a similar stimulus; later growth of the carnation ovary appears to depend on fertile seeds being present.

In _Digitalis_, pollination promotes corolla abscission without affecting its composition or turgor. But even so, it is suggested that ethylene plays an important role because pollination enhances ethylene production and abscission-zone weakening, and the latter is also caused by
exogenous ethylene (24). Similarly, an increase in ethylene production
accompanied abscission-zone weakening in the unpollinated flowers.
However, in _Musa_, non-abscising flowers do not respond to ethephon in
contrast to those which do abscind (70), which suggests that there are
flower species, or morphological types within species, in which ethylene
may have less significance.

In conclusion, ethylene appears to synchronize some of the morphological responses of flower tissues of some species at the end of natural
senescence or following pollination. The stimulus which activates the
ethylene synthesis is not known for certain; but clearly a number of
growth substances will simulate it. Further work should clarify this
problem and explain why flowers become more sensitive to ethylene as they
mature.

REFERENCES
1. Crocker W, Knight LI. 1908. Effect of illiminating gas and ethylene
   upon flowering carnations. Botanical Gazette, 46, 259-275.
2. Hasek RF, James HA, Sciaroni, RH. 1969. Ethylene - its effect on
   flower crops. Florists Review 144, 372:21, 65-68, 79-82; 3722: 16-17,
   53-56.
3. Beyer EM Jr. 1980. Ethylene and senescence in flowers. Florists Review
   165, 26-28.
4. Wang CY, Baker JE. 1980. Extending the vase life of cut flowers with
   inhibitors of ethylene synthesis and action. Florists Review 16,
   58-59.
5. Pratt HK, Goeschl JD. 1969. Physiological roles of ethylene in plants.
   Annual Review of Plant Physiology 20, 541-584.
6. Sacher JA. 1973. Senescence and post harvest physiology. Annual
   Review of Plant Physiology 24, 197-224.
7. Lieberman M. 1979. Biosynthesis and action of ethylene. Annual Review
   of Plant Physiology 30, 533-591.
8. Halevy AH, Mayak S. 1979. Senescence and post harvest physiology of
   cut flowers, Part 1. Horticultural Reviews 1, 204-236.
9. Halevy AH, Mayak S. 1981. Senescence and post harvest physiology of
   cut flowers, Part 2. Horticultural Reviews 3, 59-143.
10.Abeles FB. 1973. Ethylene in plant biology. New York and London,
   Academic Press.
11.Barden LE, Hanan JJ. 1972. Effect of ethylene on carnation keeping
   life. Journal of the American Society for Horticultural Science 97,
   785-788.

12. Camprubi P, Nichols R. 1978. Effects of ethylene on carnation flowers (Dianthus caryophyllus) cut at different stages of development. Journal of Horticultural Science 53, 17–22.

13. Camprubi P, Nichols R. 1979. Ethylene-induced growth of petals and styles in the immature carnation inflorescence. Journal of Horticultural Science 54, 225–228.

14. Mayak S, Halevy AH. 1980. Ch.7 Flower Senescence, pp. 131–156, in "Senescence in Plants" ed. KV Thimann, Boca Raton, Florida, CRC Press.

15. Davidson OW. 1949. Effects of ethylene on orchid flowers. Proceedings of the American Society For Horticultural Science, 53, 440–446.

16. Fischer CW. 1949. Snapdragons and calceolarias gas themselves. New York State Flower Growers Bulletin 52, 5–8.

17. Nichols R. 1968. The response of carnation (Dianthus caryophullus) to ethylene. Journal of Horticultural Science 43, 335–349.

18. Marousky FJ, Harbaugh BK. 1979. Ethylene-induced floret sleepiness in Kalanchoe blossfeldiana Poelln. HortScience 14, 505–507.

19. Suttle JC, Kende H. 1980. Ethylene action and loss of membrane integrity during petal senescence in Tradescantia. Plant Physiology 65, 1067–1072.

20. Kende H, Baumgartner B. 1974. Regulation of aging in flowers of Ipomoea tricolor by ethylene. Planta 116, 279–289.

21. Arditti J, Hogan NM, Chadwick AV. 1973. Post-pollination phenomena in orchid flowers. IV. Effects of ethylene. American Journal of Botany 60, 883–888.

22. Cameron AC, Reid MC. 1981. The use of silver thiosulfate anionic complex as a foliar spray to prevent flower abscission of Zygocactus. HortScience 16, 761–762.

23. Harkema H, Woltering EJ. 1981. Ethyleenschade bij snijbloemene. Vakblad voor de Bloemisterij 22, 40–42.

24. Stead AD, Moore KG. 1983. Studies on flower longevity in Digitalis. Planta 157, 15–21.

25. Dodds JH, Hall MA. 1982. Metabolism of ethylene by plants. International Review of Cytology 76, 299–355.

26. Smith WH, Meigh DF, Parker JC. 1964. Effect of damage and fungal infection on the production of ethylene by carnations. Nature 204, 92–93.

27. Mayak S, Dilley D. 1976. Regulation of senescence in carnation (Dianthus caryophyllus). Effect of abscisic acid and carbon dioxide on ethylene production. Plant Physiology 58, 663–665.

28. Ronen M, Mayak S. 1981. Interrelationship between abscisic acid and ethylene in the control of senescence processes in carnation flowers. Journal of Experimental Botany 32, 759–765.

29. Sacalis JN, Nichols R. 1980. Effects of 2:4–D uptake on petal senescence in cut carnation flowers. HortScience 15, 499–500.

30. Wulster G, Sacalis J, Janes HW. 1982. Senescence in isolated carnation petals. Effects of indoleacetic acid and inhibitors of protein synthesis. Plant Physiology 70, 1039–1043.

31. Eisinger W. 1982. Regulation of carnation flower senescence by ethylene and cytokinins. Plant Physiology 69,(4), p.136.

32. Mor Y, Spiegelstein H, Halevy AH. 1983. Inhibition of ethylene biosynthesis in carnation petals by cytokinin. Plant Physiology 71, 541–546.

33. Beyer EM. 1977. $^{14}C_2H_4$: Its incorporation and oxidation to $^{14}CO_2$ by cut carnations. Plant Physiology 60, 203–206.

34. Beyer EM, Sundin O. 1978. $^{14}C_2H_4$ metabolism in Morning Glory flowers. Plant Physiology 61, 896–899.

35. Yang SF. 1980. Regulation of ethylene biosynthesis. HortScience 15, 238-243.
36. Baker JE, Wang CY, Lieberman M, Hardenburg RE. 1977. Delay of senescence in carnations by a rhizobitoxine analog and sodium benzoate. HortScience 12, 38-39.
37. Broun R, Mayak S. 1981. Aminooxyacetic acid as an inhibitor of ethylene synthesis and senescence in carnation flowers. Scientia Horticulturae 15, 277-282.
38. Mor Y, Reid MS. 1981. Isolated petals - a useful system for studying flower senescence. Acta Horticulturae 113, 19-25.
39. Veen H, Kwakkenbos AAM. 1983. The effect of silver thiosulphate pretreatment on 1-aminocyclopropane-1-carboxylic acid content and action in cut carnations. Scientia Horticulturae 18, 277-286.
40. Beyer EMJ. 1976. A potent inhibitor of ethylene action in plants. Plant Physiology 58, 268-271.
41. Veen H, Van de Geijn SC. 1978. Mobility and ionic form of silver as related to longevity of cut carnations. Planta 140, 93-96.
42. Veen H. 1983. Silver thiosulphate: an experimental tool in plant science. Scientia Horticulturae 20, 211-224.
43. De Stigter HCM. 1981. Ethephon effects in cut °Sonia' roses after pretreatment with silver thiosulphate. Acta Horticulturae 113, 27-31.
44. Nichols R, Kofranek AM. 1982. Reversal of ethylene inhibition of tulip stem elongation by silver thiosulphate. Scientia Horticulturae 17, 71-79.
45. Mayak S, Halevy AH. 1972. Interrelationships of ethylene and abscisic acid in the control of rose petal senescence. Plant Physiology 50, 341-346.
46. Bufler G, Mor Y, Reid MS, Yang SF. 1980. Changes in 1-aminocyclopropane-1-carboxylic acid content of cut carnation flowers in relation to their senescence. Planta 150, 439-442.
47. Swart A. 1981. Quality of Lilium 'Enchantment' flowers as influenced by season and silver thiosulphate. Acta Horticulturae 113, 45-49.
48. Van Meeteren U, De Proft M. 1982. Inhibition of flower bud abscission and ethylene evolution by light and silver thiosulphate in Lilium. Physiologia Plantarum 56, 236-240.
49. Van Meeteren U. 1982. Light controlled flower bud abscission of Lilium 'Enchantment' is not mediated by photosynthesis. Acta Horticulturae 128, 37-45.
50. Kamerbeek GA, De Munk WJ. 1976. A review of ethylene effects in bulbous plants. Scientia Horticulturae 4, 101-115.
51. Veen H, Henstra S, De Bruyn WC. 1980. Ultrastructural localization of silver deposits in the receptacle cells of carnation flowers. Planta 148, 245-250.
52. Nichols R, Atkey PT. 1981. Localisation of silver deposits in flower tissues. Glasshouse Crops Research Institute Annual Report 1980. pp 41-42.
53. Adam Z, Borochov A, Mayak S, Halevy AH. 1983. Correlative changes in sucrose uptake, ATPase activity and membrane fluidity in carnation petals during senescence. Physiologia Plantarum 58, 257-262.
54. Borochov A, Halevy AH, Shinitzky M. 1982. Senescence and the fluidity of rose petal membranes. Relationship to phospholipid metabolism. Plant Physiology 69, 296-299.
55. Konze JR, Jones JF, Boller T, Kende H. 1980. Effect of 1-aminocyclopropane-1-carboxylic acid on the production of ethylene in senescing flowers of Ipomoea tricolor CAV. Plant Physiology 66, 566-571.

56. Mayak S, Legge RL, Thompson JE. 1981. Ethylene formation from 1-aminocyclopropane-1-carboxylic acid (ACC) by microsomal membranes from senescing carnation flowers. Planta 153, 49-55.

57. Thompson JE, Mayak S, Shinitzky M, Halevy AH. 1982. Acceleration of membrane senescence in cut carnation flowers by treatment with ethylene. Plant Physiology 69, 859-863.

58. Apelbaum A, Wang SY, Burgoon AC, Baker JE, Lieberman M. 1981. Inhibition of the conversion of 1-aminocyclopropane-1-carboxylic acid to ethylene by structural analogs, inhibitors of electron transfer, uncouplers of oxidative phosphorylation, and free radical scavengers. Plant Physiology 67, 74-79.

59. Wang CY, Baker JE. 1979. Vase life of cut flowers treated with rhizobitoxine analogs, sodium benzoate, and isopentenyl adenosine. HortScience 14, 59-60.

60. Leshem Y, Barness G. 1982. Lipoxygenase as effected by free radical metabolism: senescence retardation by the xanthine oxidase inhibitor allopurinol. In Biochemistry and Metabolism of Plant Lipids, 275-278, Amsterdam, Elsevier Biomedical Press.

61. Suttle JC. 1981. Effect of polyamines on ethylene production. Phytochemistry 20, 1477-1480.

62. Wang CY, Baker JE. 1980. Extending vase life of carnations with aminooxacetic acid, polyamines, EDU and CCCP. HortScience 15, 805-806.

63. Burg SP, Dijkman MJ. 1967. Ethylene and auxin participation in pollen induced fading of Vanda orchid blossoms. Plant Physiology 42, 1648-1650.

64. Strauss MS, Arditti J. 1982. Post pollination phenomena in orchid flowers. X. Transport and fate of auxin. Botanical Gazette 143, 286-293.

65. Nichols R, Bufler G, Mor Y, Fujino DW, Reid MS. 1983. Changes in ethylene production and 1-aminocyclopropane-1-carboxylic acid content of pollinated carnation flowers. Journal of Plant Growth Regulation 2, 1-8.

66. Whitehead CS, Fujino DW, Reid MS. 1983. Identification of the ethylene precursor 1-aminocyclopropane-1-carboxylic acid (ACC) in pollen. Scientia Horticulturae 21, 291-297.

67. Sacalis J, Wulster G, Janes H. 1983. Senescence in isolated carnation petals: differential response of various petal portions to ACC, and effects of uptake of exudate from excised gynoecia. Zeitschrift Fur Pflanzenphysiologie 112, 7-14.

68. Gilissen LJW. 1976. The role of the style as a sense-organ in relation to the wilting of the flower. Planta 131, 201-202.

69. Whitehead CS, Halevy AH, Reid MS. Roles of ethylene and ACC in pollination and wound-induced senescence of Petunia hybrida L. (personal communication).

70. Israeli Y, Blumenfeld A. 1980. Ethylene production by banana flowers. HortScience 15, 187-189.

ON ETHYLENE, CALCIUM AND OXIDATIVE MEDIATION OF WHOLE APPLE FRUIT
SENESCENCE BY CORE CONTROL

Y.Y. LESHEM*, I.B. FERGUSON** and S. GROSSMAN*

*Department of Life Sciences, Bar-Ilan University, Ramat-Gan 52100,
Israel, and **Division of Horticulture and Processing, DSIR, Private
Bag, Auckland, New Zealand

SUMMARY

When comparing core and flesh tissue from post-climacteric apple
(Malus domestica Linn.) fruit, it was found that core tissue manifested
considerably higher activities of lipoxygenase, and superoxide dismutase and less endogenous anti-oxidant activity, while ethylene evolution in both types of tissue was approximately equal. Tree sprays of
$Ca(NO_3)_2$ more than halved core ethylene evolution while having no
effect on flesh ethylene or on lipoxygenase in either flesh or core.
Results are interpreted in the light of a hypothesis that the more
intensive core oxidative metabolism may play a key role in senescence
of the whole fruit.

INTRODUCTION

The pome type fruit derived from an inferior ovary is composed of
two essentially discrete types of tissue. The flesh which is the major
part of the "fruit", as pointed out by Esau (1958) and Chandler (1951),
is composed of extracarpellary tissue arising from fused bases of floral
parts while the core, botanically the true fruit, is the gynoecium comprised of five folded carpels housing the seed locules. Depending on
cultivar, the boundary between the gynoecium and floral tube extracarpellary tissue may or not be discernible as a residual epidermis. As
shown by Bain and Robertson (1951) and our own preliminary microscopic
observations (Figure 1) core tissue has smaller and denser cells, contains more and smaller fat globules and possesses less amyloplasts than
the outer part of the pome.

While the core tissue is only a relatively small part of the whole
'fruit' it may nevertheless effect apple maturation by production of
senescence promoting agents which may diffuse out into the less dense

Y. Fuchs and E. Chalutz (eds.) Ethylene: Biochemical, Physiological and Applied Aspects.
ISBN 90-247-2984-X. Printed in The Netherlands
©1984, Martinus Nijhoff/Dr W. Junk Publishers, The Hague.

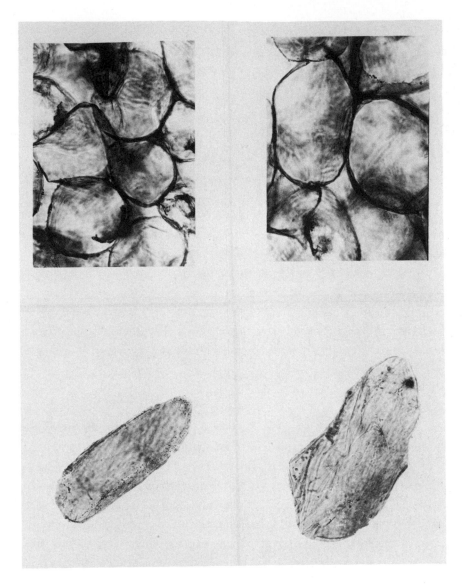

FIGURE 1. <u>Above</u>: <u>Comparison of cell size and density in post climacteric apple core (left) and flesh (right)</u>. Note greater cell density, smaller size and less intercellular cell spaces in core. Cell walls were stained with toluidene-blue-O reagent and observed under a Nikon epifluorescent microscope using phase contrast under white light.
<u>Below</u>: <u>Fat globules in core (left) and flesh (right) cells</u>. Note that the core cell contains comparatively more fat globules than flesh cells. Globules here appear as small drops along the interior cell wall periphery and possibly reflect a greater extent of bilamellar membrane breakdown. Preparations were stained with Sudan IV reagent and excess non lipid-bound stain washed out with ethanol.

flesh tissue and there contribute to whole fruit ripening. It was considered of interest to ascertain, in terms of parameters of senescence physiology, whether the anatomical differences between core and flesh are reflected in differential levels of lipoxygenase (LOX), superoxide dismutase (SOD) and ethylene production. LOX is of common occurrence in plants and attacks polyunsaturated fatty acids which contain a 1,4-cis, cis-pentadiene system, and so doing produces several species of potentially detrimental free radicals including the linoleic free radical, superoxide and indirectly, the hydroxyl free radical (DeGroot et al. 1973, and cf. review by Leshem 1981). SOD acts to dismute superoxide to molecular oxygen and its involvement in plant senescence physiology has been documented by Baker, Lieberman and Anderson (1978), McRae, Baker and Thompson (1981), Leshem et al. (1981, 1982) and others.

Pinsky, Grossman and Trop (1971) have reported that plant tissues contain water soluble, heat resistant anti-oxidants but at present, little information exists as to their content in pome fruits. In view of our interest in oxidative enzymes (LOX and SOD), and possible differences between the flesh and the more protected core, the present investigation included anti-oxidant assays. The importance of oxidative processes, active oxygen and auxin catabolism in fruit senescence has furthermore been reported by Brennan and Frenkel (1977).

Since ethylene plays a major role in fruit senescence (Lieberman 1979, Yang et al. 1982), the above observations were made concurrently with endogenous ethylene determinations on both core and flesh tissue. Also, since Ca^{2+} can be shown to delay senescence processes of both fruit (Bangerth 1979, Legge et al. 1982) and foliage (Poovaiah and Leopold 1973, Ferguson, Watkins and Harmon 1983), it was considered of interest to ascertain the degree of correlation between LOX, SOD and ethylene in both tissues of the fruit in relation to the differences between fruit sprayed or unsprayed with Ca^{2+}.

3. MATERIALS AND EXPERIMENTS

3.1. Plant material

Enzyme, antioxidant and ethylene assays were performed on the Cox's Orange Pippin, Golden Delicious or Starking cultivars. The Cox's Orange Pippin fruit were obtained from the DSIR Appleby Research Orchard, near

Nelson, South Island, New Zealand, whose trees were sprayed six times with 600 g/100 1 $Ca(NO_3)_2$ from early December to within 72 h of fruit harvest, which was mid-February. Apples were placed in polyethylene bags in cold storage at 4°C, air shipped to Israel and assayed in mid-March. The latter two cultivars were from orchards in the Golan heights in Northern Israel and were likewise stored after picking. All fruit was in the postclimacteric stage of ripening.

LOX determinations

The polarigraphic method employing O_2 electrodes utilized by Grossman and Zakut (1978) and Grossman and Leshem (1978) was followed and all determinations were carried out in quadruplicate. With the aid of a sharpened hemispherical fruit scoop, for each replicate 5 g spheres of apple tissue were taken either from the core region or from the equatorial plane of flesh tissue. Before homogenizing core tissue, seed locules were carefully removed. Homogenization was in 0.05M 1-(N morpholino) ethane sulphionic acid buffer (MES) pH 6.5 which included $5x10^{-4}M$ 2-mercaptobenzothiazole (MBT), which checks phenol oxidation but does not affect CO_2 exchange (Ferguson and Watkins, 1981). Protein was measured by Coomassie blue dye binding as outlined by Bradford (1976).

SOD determination

SOD was assayed by the cytochrome c oxidation method detailed by McCord and Fridovich (1969), which is based on the degree of superoxide dismutation caused by the plant extract applied to a superoxide generating system comprised of xanthine and xanthine oxidase. Protein was determined as in the LOX assay and number of replicates and method of sampling were also identical to the above.

Anti-oxidant determination

The method employed for anti-oxidant assay and the inhibition of linoleate oxidation by the anti-oxidant extracts were detailed by Pinsky et al. (1971) and Grossman and Zakut (1978). Aliquots, each of 0.5ml boiled anti-oxidant extracts from either core or flesh homogenate were added to a test mixture containing linoleic and commercial (Sigma) lipoxygenase and oxygen uptake was followed using the polarographic technique (Grossman and Zakut 1978). Anti-oxidant activity was expressed as specific inhibition/mg protein.

Ethylene evolution

Employing an identical sampling technique as in the LOX assays, 5g fresh wt. spheres were excised from either core or flesh tissue. These were quartered and placed in open 20ml Vacu-Cap glass vials containing 3 ml of the MBT containing MES buffer as described previously. Each vial was provided with a small test tube containing 2 ml 1M KOH as a CO_2 absorbant thus minimizing CO_2 interference on subsequent GC readings. Vials were then placed in a Tuttenauer shaking water both at 20°C for 18 h in the dark, subsequently sealed with Vacu-Caps and shaken for a further 3 h. With a hypodermic syringe, 3 ml gas samples were drawn from each vial and ethylene assayed on a Packard Model 896 Flame Ionization Detection Gas Chromatograph using a 0.5 ppm ethylene standard, attenuation being $3x10^{-12}$. All tests were in quadruplicate. All results were subjected to statistical 'Analysis of Variance' and probability (p) of significance was calculated at the p $<0.05\%$ level. Standard Deviations were calculated for results presented in Fig. 1.

RESULTS

LOX, SOD, anti-oxidants and ethylene

A series of comparisons of the above criteria was carried out in all of the three apple cultivars investigated. Results presented in Table 1

Table 1. Comparison of LOX, SOD, anti-oxidant and ethylene levels in apple core and flesh tissue.

Results are of 4 replicates and represent one of a typical series in the Cox's Orange Pippin cultivar. Values entered are functions of 1 mg protein. See 'Methods' for experimental details.

Tissue	LOX μl O_2/min	SOD Units	Anti-oxidants Specific inhibition	Ethylene ppm
Core	17.7	15.8	728	0.33
Flesh	5.8	6.2	1075	0.28
Level of Statistical Significance-p	<0.01	<0.01	<0.05	n.s.*

*Non significant

for Cox's Orange Pippin are typical of several trials all of which clearly indicated similar patterns. From these results it is apparent that core tissue manifests significantly higher activity of both LOX and SOD, but somewhat less anti-oxidant activity. The small difference observed in ethylene between the two types of tissue is not significant.

Effect of Ca^{2+} on endogenous levels of ethylene and LOX in apple core and flesh

This experiment was performed only on the Cox's Orange Pippin culti-var the fruit of which, as previously stated, originated from Ca^{2+}-treated orchard trees in Nelson, New Zealand. In this series, anti-oxidants and SOD were not assayed. As seen in the results (Fig. 2),

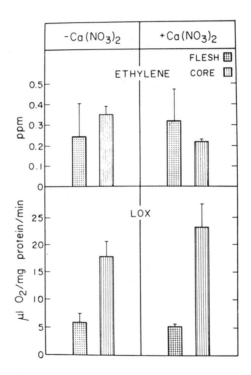

FIGURE 2. Effect of Ca^{2+} treatment on ethylene production and LOX levels in Cox's Orange Pippin apple flesh and core.
Note: $Ca(NO_3)_2$ was spray applied to orchard trees. See 'Methods' for experimental details. Vertical lines on column heads indicate Standard Deviations.

Ca^{2+} spray had no significant effect on LOX activity in core and flesh. However, ethylene production from core tissue of Ca^{2+}-treated fruit was only half that from untreated fruit. There were no significant differences between the flesh samples, and as shown in Table 1 above, core tissue in untreated fruit had higher ethylene production than flesh tissue.

DISCUSSION

As seen in Table 1, anatomical differences between core and flesh tissue outlined in the Introduction and observations in Fig. 1 are also expressed in enzymatic patterns. It is apparent that in this late stage of fruit ripening, the core or "true fruit" tissue significantly exceeds flesh or extracarpellary tissue in endogenous content of the two assayed enzymes - LOX and SOD - both of which have been reported to be associated with induction of senescence (Grossman and Leshem 1978, Baker et al. 1978, Leshem et al. 1982). Concerning ethylene evolution, the same trend was observed but this was not statistically significant. These results are a little unexpected since both LOX and SOD are connected with (per)oxidative metabolism which in the core, being situated farthest away from the external atmosphere, should presumably be active to a lesser degree. However, the higher cell density, and the non-storage (less amyloplasts) function of the pome gynoecium possibly involves a higher cellular metabolic turnover which, in turn, leads to enhanced LOX and SOD activity. Moreover in support to this assumption DeGroot et al. (1973) have indicated that under conditions tending to the anaerobic, linoleic free radicals are formed.

It is suggested that while LOX may function to produce free radicals, the SOD increment acts to scavenge free radicals already produced. This implies, as suggested by Fridovich (1981) that SOD is an inducive enzyme. The core, moreover, is equipped with somewhat less anti-oxidant (Table 1) capacity, thus concomitantly more LOX activity could be manifested. This observation is possibly related to the observation in most apple cultivars that phenolic browning in freshly cut and air-exposed tissue of most apple cultivars is more pronounced in core than in flesh tissue. In this context, Feys et al. (1980) have reported that LOX activity in apple core is comparatively high, increases upon storage and precedes core browning.

Based upon the above observations, it may follow that senescence of the whole apple fruit is at least partly linked to the more intensive core oxidative enzymic activity which could promote or initiate processes of maturation. The ethylene measurements reported here are probably too late in the ripening process to show differences in the synthetic capacity of the two tissues. However, the reduction in core ethylene found with Ca^{2+} spray treatment does indicate that the core and flesh may potentially have different responses to the induction of ethylene synthesis. Ca^{2+}-sprayed fruit is generally less mature (C.B. Watkins, unpublished) and this may be reflected in the lower ethylene rates shown in Figure 2.

These results do raise the question of whether Ca^{2+} treatments affect the overall ripening of the fruit, or the fruit metabolism at localized sites. Vacuum infiltration and spray treatments with Ca^{2+} do result in higher flesh levels, but often have little effect on core Ca^{2+} content (Ferguson and Watkins 1983). A closer examination of metabolic activity of the various tissues of the fruit in relation to ripening, calmodulin and Ca^{2+} effects should prove useful.

Acknowledgements - The authors are deeply indebted to Dr. Yoram Fuchs, of the Division of Storage, Volcani Research Centre, Israel, for his advice and aid in the ethylene determinations, to Dr. Yehudit Wurzburger of Bar-Ilan University for her aid with the SOD assays, to Dr. C. Peterson and Gregory Moon of the Department of Biology, University of Waterloo, Canada for microscopic sectioning.

This research was in part made possible by assistance from the Division of Horticulture and Processing, DSIR, Auckland, New Zealand, to Y.L.

REFERENCES

Bain J, Robertson RN. 1951. The physiology of growth in apple fruits: Cell size, cell number and fruit development. Austral. J. Sci. Res. B 4: 75-91.
Baker JE, Lieberman M, Anderson JD. 1978. Inhibition of ethylene production on fruit slices by a rhizobitoxine analog and free radical scavengers. Plant Physiol. 61: 886-888.

Bangerth F. 1979. Calcium-related disorders of plants. Ann. Rev. Phytopathol. 17: 97-122.

Bradford M. 1976. A rapid and sensitive method for the quantitation of microgram quantities of protein utilizing the principle of dye binding. Anal. Biochem. 72: 248-254.

Brennan T, Frenkel C. 1977. Involvement of hydrogen peroxide in the regulation of senescence in pear. Plant Physiol. 59: 411-416.

Chandler WH. 1951. Deciduous Orchards. Lea and Fabiger, Philadelphia,

DeGroot JJMC, Garsen GJ, Vliegenthart JFG, Boldingh J. 1973. The detection of linoleic acid radicals on the anaerobic reaction of lipoxygenase. Biochim. Biophys. Acta 326: 279-284.

Esau K. 1958. Plant Anatomy. Wiley and Sons, New York.

Ferguson IB, Watkins CB. 1981. Ion relations of apple fruit tissue during fruit development and ripening. III. Calcium uptake. Austral. J. Plant Physiol. 8: 259-266.

Ferguson IB, Watkins CB. 1983. Cation distribution and balance in apple fruit in relation to calcium treatments for bitter pit. Sci. Hortic. (in press).

Ferguson IB, Watkins CB, Harman JE. 1983. Inhibition by calcium of senescence of detached cucumber cotyledons: effect on ethylene and hydroperoxide production. Plant Physiol. 79: 182-186.

Feys M, Naesens W, Tobback P, Maes E. 1980. Lipoxygenase activity in apples in relation to storage and physiological disorders. Phytochemistry 19: 1009-1011.

Fridovich I. 1981. The biology of superoxide and of superoxide dismutases - in brief. In: Oxygen and Oxy-Radicals in Chemistry and Biology (Rodgers M., Powers E, eds.), New York, Academic Press, pp. 197-204.

Grossman S, Leshem, Y. 1978. Lowering of endogenous lipoxygenase activity Pisum sativum foliage by cytokinin as related to senescence. Physiol. Plant. 43: 359-362.

Grossman S, Zakut, R. 1978. Determination of lipoxygenase activity. In: Methods of Biochemical Analysis (Glick D, ed.), Vol. 25, New York, Wiley & Sons, pp. 305-329.

Legge RL, Thompson JE, Baker JC, Lieberman M. 1982. The effect of calcium on the fluidity and phase properties of microsomal membranes isolated from post-climacteric Golden Delicious apples. Pl. Cell Physiol. 23: 161-170.

Leshem Y. 1981. Oxy free radicals and plant senescence. What's New in Plant Phys. 12: 1-4.

Leshem Y, Wurzburger J, Grossman S, Frimer A. 1981. Cytokinin interaction with free radical metabolism and senescence: effects of endogenous lipoxygenase and purine oxidation. Physiol. Plant. 53: 9-12.

Leshem Y, Wurzburger J, Frimer AA, Barness G, Ferguson IB. (1982). Calcium and calmodulin metabolism in senescence: interaction of lipoxygenase and superoxide dismutase with ethylene and cytokinin. In: Plant Growth Substances, 1982. Proc. 11th IPGSA Conference, Aberstwyth, UK. (Wareing PF, ed.), New York, Academic Press, pp. 569-578.

Lieberman M. 1979. Biosynthesis and action of ethylene. Ann. Rev. Plant Physiol. 30: 533-591.

McCord JM, Fridovich I. 1969. Superoxide dismutase: an enzymic function for erythrocuprein (hemocuprein). J. Biol. Chem. 244: 6049-6055.

McRae DG, Baker JE, Thompson JE. 1982. Evidence for involvement of the superoxide radical in the conversion of 1-amino-cyclopropane acid to ethylene by pea microsomal membranes. Plant Cell Physiol. 23: 375-383.

Pinsky A, Grossman S, Trop M. 1971. Lipoxygenase content and antioxidant activity of some fruits and vegetables. J. Food Sci. 36: 571-572.

Poovaiah BW, Leopold AC. 1973. Deferral of leaf senescence with calcium. Plant Physiol. 52: 236-239.

Yang SF, Hoffman NE, McKeon T, Riov J, Kao CH, Yung KH. (1983). Mechanism and regulation of ethylene biosynthesis. In: Plant Growth Substances, 1982. Proc. 11th IPGSA Conference, Aberstwyth, UK. (Wareing PF, ed.), New York, Academic Press, pp. 234-248.

ETHYLENE-MEDIATED GROWTH RESPONSE IN SUBMERGED DEEP-WATER RICE*

H. KENDE, J.-P. METRAUX, I. RASKIN

MSU-DOE Plant Research Laboratory, Michigan State University, East Lansing, MI 48824, USA

Floating or deep-water rice is mainly grown in the floodplains of Southeast Asia where the water can rise up to 6 m during the rainy season (1). Deep-water rice has great agronomic importance because it is a subsistence crop in many densely populated areas where no other crop can be grown. The distinguishing characteristic of this rice is its ability to elongate with rising waters after 4-6 weeks of growth. The survival of the deep-water rice plant depends on its ability to keep part of its foliage above the water surface. Completely submerged plants cease to elongate and eventually die (2). When the lower leaves of deep-water rice become submerged, their growth also stops and they eventually die so that the basal part of the partially submerged plant consists of elongated internodes only. The total height of deep-water rice plants can reach up to 7 m in areas with high flood levels (3). This remarkable elongation occurs in spite of severely restricted supply of O_2 and CO_2 under water.

Of all rice varieties, deep-water rice has been studied least. Practically nothing has been known about the physiological basis of internode elongation induced by partial flooding in spite of the great potential of increasing grain yields in flooded areas by enhancing the elongation capacity of high-yielding rice varieties which otherwise cannot survive rapid flooding.

*Supported by the National Science Foundation through Grant No. PCM 8109764 and the Department of Energy under Contract No. DE-AC02-76ER01338.

Y. Fuchs and E. Chalutz (eds.) Ethylene: Biochemical, Physiological and Applied Aspects.
ISBN 90-247-2984-X. Printed in The Netherlands
©1984, Martinus Nijhoff/Dr W. Junk Publishers, The Hague.

This paper summarizes recent results from our laboratory on the physiological mechanisms involved in regulation of growth of partially flooded deep-water rice. It also discusses the similarities of growth regulation in etiolated rice seedlings and deep-water rice plants.

THE EFFECT OF SUBMERGENCE AND ETHYLENE ON THE GROWTH OF DEEP-WATER RICE PLANTS.

In our initial experiments, we investigated the effects of submergence and ethylene treatment on internodal growth of intact plants (4). Up to the age of 21 days, 7 days of submergence did not stimulate internodal elongation. After that time, a 2- to 5-fold increase in total internodal length was obtained when plants were submerged for 7 days. Total internodal growth during that period was mainly based on enhanced elongation of existing internodes rather than on an increase in the number of internodes. When 35-day-old plants were immersed to a depth of 60 cm and then lowered, over a 6-day period, each day 10 cm deeper into the water, internodal growth was stimulated within less than 24 h. During the same time, the ethylene concentration in the lacunae began to rise from the initial level of 0.02 μl/l reaching a value of 1 μl/l after 48 h of submergence. It remained constant at this level for the remainder of the experiment.

When ethylene at 0.4 μl/l was applied to non-submerged plants, total internodal elongation as well as elongation of individual internodes was enhanced 2- to 3-fold. The growth response of non-submerged plants was saturated at 0.05 μl/l of ethylene. Floating rice plants had to be at least 28 days old before internodal elongation could be stimulated by treatment with ethylene.

The role of endogenous ethylene in partially submerged plants was examined by inhibiting ethylene biosynthesis using aminooxyacetic acid (AOA) and aminoethoxyvinylglycine (AVG). In order to reduce ethylene formation to as low a level as possible, AOA (10^{-4} M) was added to the water in which the plants were submerged, and AVG (10^{-3} M) was injected into the lacunae in agar from which it would be released slowly during the subsequent 2 days of submergence. As a result of the combined treatment with AOA and AVG, a marked inhibition of internodal elongation occurred. When 1-amino-cyclopropane-1-carboxylic acid (ACC) was added to AOA in the submergence medium and injected, together with AVG, into the internodal lacunae, internodal elongation was stimulated to a similar extent as in submerged plants injected with agar only.

We tested whether the rise in the endogenous ethylene concentration during submergence was due, at least in part, to increased ethylene synthesis or whether it was based solely on accumulation of ethylene inside the submerged tissue. We found that the rate of ethylene synthesis in sections excised from internodes of submerged plants was around ten times higher than that in sections from control plants. To distinguish between release of ethylene, which had accumulated during submergence, and enhanced ethylene synthesis, ethylene evolution was monitored at 25°C and 4°C. At the lower temperature, ethylene evolution was reduced 92% indicating that ethylene synthesis had been inhibited. AVG was also used to differentiate between release of accumulated ethylene and synthesis of new ethylene. When internode sections were incubated in 10^{-5} M AVG, ethylene production was inhibited by 90%. Therefore, release of accumulated ethylene contributed little to the total amount of ethylene evolved by internode sections from submerged plants.

REGULATION OF GROWTH IN ISOLATED STEM SECTIONS OF DEEP-WATER RICE

Isolated stem sections provided us with a system of reduced complexity and increased versatility to study the submergence response in deep-water rice (5). We evaluated especially the contribution of low O_2 and high CO_2 levels within the submerged tissue on stimulation of growth. Stem sections, 20 cm long and containing the top-most internode

124

and the highest two nodes, were excised from the main culms and tillers. These sections were either submerged in volumetric cylinders filled with water or were placed in cylinders through which the respective gas mixture was circulated. When stem sections isolated from deep-water rice plants were incubated in a stream of air or were submerged in water under continuous light for 3 days, the final length of the sections turned out to be very similar. When the stems were slit open, it became evident that the internodes of the non-submerged sections had elongated very little and that growth of the section was based mainly on elongation of the leaf sheaths and leaf blades. In contrast, the internodes of the submerged sections had increased several fold in length while leaf growth was inhibited.

In submerged sections, the lacunae of the internodes contained about 3% O_2, 6% CO_2 and 1 μl/l ethylene. Air and different mixtures of O_2 (3% or 21%), CO_2 (0.03% or 6%) and ethylene (none or 1 μl/l) in N_2 were used to evaluate the effect of high concentrations of CO_2 and ethylene and low concentrations of O_2 on growth of internodes and leaves. Ethylene at 1 μl/l in air and 3% O_2 enhanced internodal growth by 7 and 4.5 times, respectively. When 1 μl/l ethylene and 3% O_2 were supplied in the same gas mixture, internodal growth was enhanced tenfold. High concentrations of CO_2 alone had very little effect on internodal elongation in non-submerged sections. However, when 1 μl/l ethylene was added to gas mixture containing 6% CO_2, the growth of internodes was increased over ninefold. While internodal elongation was strongly promoted by ethylene, leaf growth was inhibited when ethylene was added to air or to any of the other gas mixtures. The effect of submergence on growth of internodes and leaves in deep-water rice stem sections was closely mimicked by passing a gas mixture of 3% O_2, 6% CO_2 and 1 μl/l ethylene in N_2 through the chambers containing rice stem sections.

We investigated the effect of submergence and altered gas composition on ethylene synthesis in rice stem sections.

Since it is not possible to compare ethylene evolution in submerged and non-submerged sections directly, we transferred the internodal tissue from submerged stem sections into test tubes containing 3% O_2 and 6% CO_2, concentrations that were similar to those found in the internodal lacunae of submerged sections, and determined ethylene evolution under these conditions. Internodal tissue from sections that had been incubated in air were transferred to test tubes containing air. Internodal tissue from submerged sections evolved over 5 times more ethylene than internodal tissue from air-incubated sections. We also examined the effect of low O_2 and high CO_2 concentrations on ethylene synthesis in internodal and leaf tissue of stem sections that had been incubated previously in different gas mixtures. Low O_2 levels stimulated ethylene synthesis in internodal tissue fourfold. The reverse was true for leaves. These evolved the largest amount of ethylene in air and the least in gas mixtures containing low levels of O_2. CO_2 did not significantly affect ethylene synthesis.

THE CELLULAR BASIS OF THE ELONGATION RESPONSE IN DEEP-WATER RICE

We analyzed the cellular processes underlying internodal elongation in intact rice plants and excised stem sections (6). In whole plants, submergence and exposure to ethylene led to an up to tenfold increase in the number of cells in the zones which elongated between day 0 and 3 of the respective treatment. The average cell length increased 1.6 times as a result of ethylene treatment and 1.5 times as a result of submergence. The average cell length in the newly elongated region of control internodes was similar to that of the adjacent region which had been present before the start of the experiment. We also examined the growth response of excised stem sections after submergence, exposure to ethylene and to gas mixtures of 3% O_2, 6% CO_2 in N_2 and 3% O_2, 6% CO_2, 1 $\mu l/l$ ethylene in N_2. The latter gas mixture approximates to the internal gas composition of

submerged tissue. In internodes of stem sections, we observed the same distribution of growth as in internodes of
intact plants. In all treatments, both the number of cells
and the size of cells increased in the elongating zone.

Cell divisions were localized by labeling the nuclear
DNA of dividing cells with [^{3}H]thymidine. When internodes
of submerged rice stem sections were labeled for increasing
periods of time, the zone of labeled nuclei gradually extended acropetally. The length of the intercalary meristem
could be estimated by subtracting the length of internode
added during labeling from the length of the region containing labeled nuclei. Such calculation yielded an average
meristem length of 1.8 $\pm$ 0.6 mm (n = 5, $\pm$ standard deviation). In control internodes, incorporation of [^{3}H]thymi-
dine was much lower than in submerged ones. This reflected
the lower cell-division activity in the intercalary meristem. The length of the intercalary meristem was 0.9 $\pm$ 0.1
mm (n = 3, $\pm$ standard deviation).

REGULATION OF GROWTH IN RICE COLEOPTILES

A study of the regulation of growth in rice coleoptiles
revealed several similarities to the regulation of growth in
deep-water rice plants (7). Growth of the rice coleoptile
is stimulated under water and in a closed atmosphere (for a
review, see ref. 7). Ku et al. (8) were the first to show
that ethylene promoted growth of the rice coleoptile. We
evaluated the roles of reduced O_2 levels and increased CO_2
and ethylene concentrations in the growth response of etiolated rice seedlings. Etiolated rice seedlings grown in
aerated or sealed containers exhibited marked differences in
their growth habit. The altered composition of gases in a
closed atmosphere stimulated growth of the coleoptile and
the mesocotyl and inhibited growth of the leaves. The levels of O_2, CO_2 and ethylene inside the sealed containers
were 3%, 21% and 0.9 μl/l, respectively, after 8 days of
incubation of seedlings. Different mixtures of O_2, CO_2
and ethylene in N_2 were used to evaluate the contribution

of each of these gases to the growth response. Enclosure increased coleoptile growth by 160%. High CO_2 (15%), low O_2 (3%) and ethylene (1 μl/l) applied individually in the gas stream were responsible for about one-third of the total stimulation of coleoptile growth observed in sealed containers. Combined treatment with any two of these gases elicited about two-thirds of the response caused by enclosure. The combination of ethylene, high CO_2 and low O_2 mimicked closely the effect of enclosure on rice seedlings. Neither high levels of CO_2 nor low levels of O_2 stimulated ethylene synthesis in etiolated rice seedlings. The highest rates of ethylene release were observed in seedlings grown in air.

CONCLUSIONS

Deep-water rice is adapted to grow very rapidly when partially submerged in water. This adaptation is based on the ability of deep-water rice to react to reduced O_2 and elevated CO_2 and ethylene levels within the submerged stems. Oxygen at concentrations found in submerged internodes (ca. 3%) enhances ethylene synthesis in internodal tissue. Ethylene accumulates in submerged stems and promotes internodal growth. The stimulatory effect of ethylene on internodal growth is enhanced at elevated CO_2 levels. Increased production of new cells in the intercalary meristem and their subsequent elongation form the basis of the growth response to submergence and ethylene treatment in deep-water rice plants.

Etiolated rice seedlings also respond to altered gas composition in the atmosphere. Low levels of O_2 and high concentrations of CO_2 and ethylene all induce growth. CO_2 and O_2 appear to act independently of ethylene.

REFERENCES
1. DeDatta, S.K. 1981. Principles and practices of rice production. Wiley and Sons, New York.
2. Sugawara, T., Horikawa, T. 1971. Studies on the elongation of internodes in floating rice plants - morphological and histological observations. Bull. Coll. Agric. Utsunomiya Univ. 8:25-46.

3. Vergara, B.S., Jackson, B., DeDatta, S.K. 1976.
   Deep-water rice and its response to deep-water stress.
   In: Climate and rice. International Rice Research
   Institute, Los Baños, Philippines, pp. 301-319.
4. Métraux, J.-P., Kende, H. 1983. The role of ethylene in
   the growth response of submerged deep-water rice. Plant
   Physiol. 72:441-446.
5. Raskin, I., Kende, H. 1983. Regulation of growth in stem
   sections of deep-water rice. Planta (in press).
6. Métraux, J.-P., Kende, H. 1983. The cellular basis of
   the elongation response in submerged deep-water rice.
   Planta (in press).
7. Raskin, I., Kende, H. 1983. Regulation of growth in rice
   seedlings. J. Plant Growth Regul. (in press).
8. Ku, H.S., Suge, H., Rappaport, L., Pratt, H.K. 1970.
   Stimulation of rice coleoptile growth by ethylene.
   Planta 90:333-339.

CONTROL OF THE BIOSYNTHESIS OF ETHYLENE IN SENESCING TISSUES[1,2]

Nehemia Aharoni, Sonia Philosoph-Hadas and Shimon Meir[3]
Agricultural Research Organization
The Volcani Center
Bet Dagan, Israel

INTRODUCTION

Vegetative tissues usually produce very small amounts of ethylene (2). However, stress and wounding effects, as well as application of some growth regulators, especially auxins, can cause a remarkable increase in ethylene production (1,10). The rate of ethylene production as positively correlated to the internal level of free IAA, has been demonstrated so far in mung bean hypocotyls (9) and in pea epicotyls (7). Indeed, in these two tissues, IAA-induced ethylene production continued as long as free IAA was present in the medium. On the other hand, in tobacco leaves, IAA-induced ethylene production lasted for several days (3), although IAA was removed from the medium after a 4h-pulse (unpublished results). Such a prolonged increase of the IAA-induced ethylene production occurred in tobacco leaves only when sucrose, or some other carbohydrates, were present in the incubation medium (3,12). These observations, as well as others, led us to the hypothesis that the natural regulation of ethylene biosynthesis in the leaf is primarily dependent on the type of the endogenous IAA conjugates formed in the tissue and their susceptibility to hydrolytic enzymes. This enzyme-catalyzed hydrolysis of the IAA conjugates enables a frequent, renewed supply of free IAA (3-5), which, in turn, controls ethylene production in vegetative tissues (13). In the

---

[1]Dedicated to the memory of Dr. Morris Lieberman, with whom some parts of this study were initiated.

[2]This research was supported by Grant No. I-145-79 from BARD - The United States-Israel Binational Agricultural Research and Development Fund.

[3]The authors wish to thank Ms. Orit Dvir for her skillful technical assistance.

Y. Fuchs and E. Chalutz (eds.) Ethylene: Biochemical, Physiological and Applied Aspects.
ISBN 90-247-2984-X. Printed in The Netherlands
©1984, Martinus Nijhoff/Dr W. Junk Publishers, The Hague.

present paper we summarize our recent results (3,11,12) regarding the stimulatory effect of carbohydrates both on IAA metabolism and on ethylene biosynthesis.

PROCEDURES

Experiments were carried out with discs taken from fully expanded, mature leaves of tobacco, tomato, bean and cotton, grown either outdoors or in a greenhouse, as well as with etiolated mung bean hypocotyls. Leaves were washed with tap water and surface-sterilized with sodium hypochlorite as previously described (2). The mung bean hypocotyls were grown and treated as outlined by Lau and Yang (9). All subsequent handling of the tissues involved sterile techniques. Leaf discs were incubated abaxial surface down on a filter paper in 50-ml Erlenmeyer flasks, containing 2 ml of 50 mM Na-phosphate buffer (pH 6.1) and 50 μg/ml chloramphenicol. When indicated, 0.1 mM IAA, 1 μCi $(1-^{14}C)$ IAA (59 mCi/mmol), 0.1 mM IAA-L-alanine (IAAlà), 50 mM sucrose, or 0.1 mM 1-amino-cyclopropane-1-carboxylic acid (ACC) plus 0.1 mM aminoethoxyvinylglycine (2-amino-4-(2'-aminoethoxy)-trans-3-butenoic acid (AVG), were added to the medium. Mung bean hypocotyls were incubated as described above, but without filter paper and with 3 ml incubation liquid. Two plastic center wells, hung in each flask, contained filter papers wick wetted one with 0.1 ml of 0.25 M $Hg(ClO_4)_2$ reagent for ethylene absorption and the second with 0.1 ml of 10.% KOH for CO_2 absorption. The flasks were sealed with rubber serum caps, incubated in darkness at $30^{\circ}C$ and ethylene production was assayed periodically. Where indicated, the KOH solution was omitted and an atmosphere of 10 or 15% CO_2 was obtained by injecting different amounts of pure CO_2 into each sealed flask. The ethylene absorbed to the $Hg(ClO_4)_2$ solution in the center wells, was released in sealed new Erlenmeyer flasks by injecting saturated LiCl, and measured by gas chromatography. After sampling for ethylene determination, the $Hg(ClO_4)_2$ and KOH solutions were renewed for another incubation period. For CO_2 determination, the CO_2 trapped in the KOH solution was released by acidification with lactic acid and analyzed in a gas chromatograph. Extraction and chromatography of the IAA conjugates, formed in the tissue after application of $(1-^{14}C)$ IAA, were performed as described by Aharoni and Yang (3).

RESULTS AND DISCUSSION

Stimulatory effect of sucrose on IAA and IAAla-induced ethylene production: IAA and sucrose were applied individually and simultaneously
to the leaf discs for examination of possible interactions between them.
The effect of a 10% CO_2 atmosphere was also tested. Rates of ethylene
production with these treatments were measured in the course of a 6-day
incubation (Fig. 1).

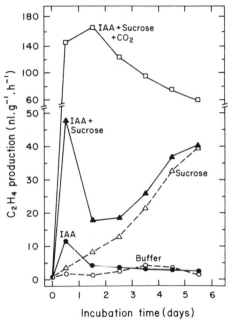

FIGURE 1. Stimulatory effects
of IAA, sucrose and CO_2 on ethylene production rates in tobacco
leaf discs.

In untreated discs the level of ethylene production was very low, and it
rose in a climacteric-like pattern on the third day. Application of IAA
at 0.1 mM, stimulated ethylene production only during the first day, and
this stimulation could be further increased by employing 50 mM sucrose.
In the combined treatment of IAA plus sucrose, a second rise of ethylene
production, which lasted for the next 4 days, could be observed. Sucrose
by itself caused a gradual increase in ethylene production, which lasted
for 6 days of incubation. When the atmosphere of the IAA-plus-sucrose-
treated discs was enriched with 10% CO_2, a great increase in ethylene production was observed. We assume that in discs treated with IAA plus
sucrose, the first short burst of ethylene production during the initial
24 h, is a result of an interaction between sucrose and the free IAA taken

132

up from the medium, whereas the second and prolonged rise results from an interaction between sucrose and the endogenous IAA conjugates. This is possible since we found that 96% of the endogenous auxins in fresh tobacco leaves appear as IAA conjugates, while only 4% remains as free IAA (11).

In order to test this postulated interaction between sucrose and IAA conjugates, IAAla, a synthetic IAA conjugate, was employed. Thus, application of 0.1 mM IAAla to tobacco leaf discs increased only slightly ethylene production, especially during the climacteric-like rise (Fig. 2).

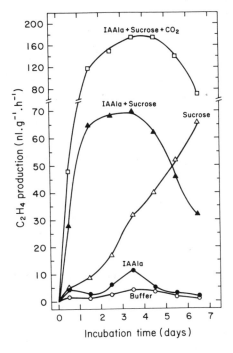

FIGURE 2. Stimulatory effects of IAAla, sucrose and CO_2 on ethylene production rates in tobacco leaf discs.

However, the combined treatment of IAAla plus sucrose caused an immediate increase in ethylene production, which lasted for 7 days. Similarly to Figure 1, also here, CO_2 markedly increased the rate of ethylene production in leaf discs treated with IAAla plus sucrose.

For elucidating the nature of the interaction between sucrose and either IAA or IAAla, their effect in inducing ethylene production was studied in different plant tissues. No stimulatory effect of sucrose was found in the following tissues, which did not respond to added IAA in increased ethylene production: apple plugs, grapefruit albedo and leaf discs of citrus, sugar beet and corn. The responsive vegetative tissues

are listed in Table 1 in descending order, according to their response to
added sucrose in the presence of either IAA or IAAla.

TABLE 1.   Effect of sucrose on ethylene production in IAA or IAAla-treated
vegetative tissues

| Plant System | Sucrose | Treatment | | | |
| | | IAA | | IAAla | |
		−Sucrose	+Sucrose	−Sucrose	+Sucrose
		--- % of control----			
Tobacco leaf discs	495	250	2114	261	2411
Tomato leaf discs	395	138	505	120	595
Cotton leaf discs	186	163	461	116	265
Bean leaf discs	156	240	346	245	218
Mung bean hypocotyls	100	300	280	276	170

IAA - 0.1 mM; IAAla - 0.1 mM; Sucrose - 50 mM.

The stimulatory effect of sucrose on ethylene production was the greatest
in tobacco leaf discs, giving a 5-fold enhancement when applied alone, a
21-fold in combination with IAA, and a 24-fold in combination with IAAla.
Tomato and cotton leaf discs were responsive as well. In bean leaves,
the sucrose had a slight effect when applied alone or with IAA, but was
ineffective when applied with IAAla. Mung bean hypocotyls were not responsive at all to added sucrose, although they showed an IAA or IAAla-
induced ethylene production.

The results may suggest that the sucrose stimulating effect is dependent on the presence of IAA conjuates in the tissue, and on their capability to undergo enzymatic hydrolysis (3,11).

The composition of IAA conjugates in IAA-treated tobacco leaf discs
and mung bean hypocotyls: The ineffectiveness of added sucrose to
stimulate ethylene production in mung bean hypocotyls (Table 1), could be
related to a different IAA metabolism in this tissue. Figures 3 and 4
illustrate the pattern of IAA conjugates formed in tobacco leaf discs or
in mung bean hypocotyls, respectively, during 24 h of incubation with
$(1-^{14}C)$IAA. The ethanolic extracts of these tissues were chromatographed
on paper using 1-butanol-acetic acid-H_2O (4:1:4, v/v), and the chromatograms were then radio scanned. The three IAA metabolites of tobacco leaf
shown in Figure 3 have been identified as IAA conjugates, two of which
were characterized as an ester-linked IAA (conjugate 2), and a peptide-
linked IAA (conjugate 3), apparently not indole-3-acetyl-aspartic acid

134

(IAAsp) (11). When eluted from the chromatograms and reapplied to tobacco leaf discs, all these 3 IAA conjugates could stimulate ethylene production (3,11), and sucrose enhanced considerably their hydrolysis, especially that of conjugate 2, the esteric IAA (11). Parallelly, we have also found that an ester-linked IAA is the major component (84%) of the naturally occurring IAA conjugates in the tobacco leaf (11).

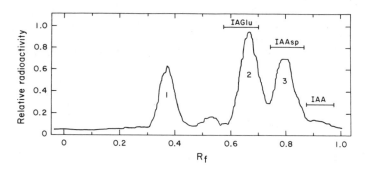

FIGURE 3. Radiochromatogram scan of extract from tobacco leaf discs incubated with (1-^{14}C)IAA for 24 hours.

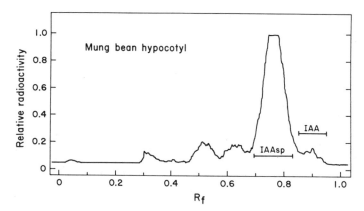

FIGURE 4. Radiochromatogram scan of extract from mung bean hypocotyls incubated with (1-^{14}C)IAA for 24 hours.

When (1-^{14}C)IAA was applied to mung bean hypocotyls, only IAAsp was accumulated (Fig. 4). This IAA conjugate was almost ineffective in stimulating ethylene production both in pea stems (6) and in tobacco leaf discs (unpublished results). Thus, it seems that unlike with the tobacco IAA conjugates (Fig. 3), sucrose has almost no effect on hydrolysis of IAAsp. Persumably, this may be one of the reasons for the failure of

added sucrose to stimulate ethylene production in mung bean hypocotyls
(Table 1), which contain IAAsp as their main IAA conjugate (Fig. 4).

Possible mechanisms for the carbohydrates-stimulated ethylene production: In addition to the presumed IAA-mediation in the sucrose-stimulated ethylene production on the step of ACC formation, Figures 1 and 2
suggest that CO_2 is also involved (1), but in the step converting ACC to
ethylene (Fig. 5 and ref. 8).

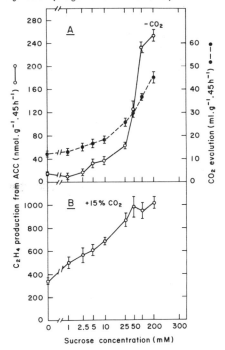

FIGURE 5. Effects of sucrose concentrations on ACC-dependent ethylene production and on respiratory
CO_2 in tobacco leaf discs.

The results of Figure 5A imply that most of the sucrose stimulation on
ACC-dependent ethylene production can be attributed to the CO_2 produced
by the leaf discs, according to the following mechanism: sucrose enhances
respiratory CO_2 (Fig. 5A), which, in turn, acts synergistically with the
sucrose present in the tissue, thereby causing a further enhancement of
ACC-dependent ethylene production (Figs. 5A,5B). It should be noted that
the most remarkable synergistic effect between CO_2 and sucrose was obtained with very low sucrose concentrations, which may represent the physiological situation in leaves held in darkness. Thus, simultaneous application of only 1 mM sucrose with 15% CO_2 to the leaf discs, gave a 50-fold
increase in ethylene production. Additionally, regarding the 3-fold enhancement of ACC-dependent ethylene production obtained between 1 to 50 mM

136

sucrose in the presence of saturating 15% CO_2 (Fig. 5B), it is plausible that also sucrose by itself can directly promote this step. However, it seems that most of the sucrose-stimulated ethylene production on the site of ACC conversion to ethylene, can be mainly ascribed to the sugar-stimulated endogenous CO_2, which acts in synergism with sugars in the tissue.

The scheme illustrated in Figure 6 summarizes the involvement of sugars in auxin metabolism and in the pathway of ethylene biosynthesis as follows: (a) Sugars stimulate IAA uptake in the model system of tobacco leaf discs; (b) Sugars enhance the hydrolysis of some IAA conjugates and this is accompanied by increased oxidation of the released free IAA; (c) The free IAA released induces ACC formation from s-adenosylmethionine (SAM), the last two intermediates in the pathway of ethylene biosynthesis (13); and (d) Sugars stimulate the respiratory CO_2 (Fig. 5A), which, in turn, stimulates ethylene production (Figs. 1,2) by enhancing ACC conversion to ethylene (8) (Fig. 5B). Evidence for these postulated mechanisms has been reported recently (3,11).

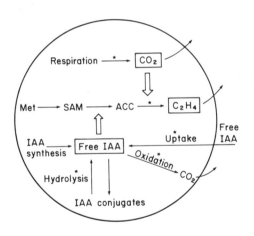

FIGURE 6. Involvement of carbohydrates in auxin metabolism and ethylene biosynthesis pathway of tobacco leaf discs.

*Asterisks indicate sites of carbohydrates' action.

Ethylene-mediated effect of sucrose in cotton plants: So far, all the experiments reported herein, were performed with leaf discs or other wounded tissues. However, in preliminary experiments, we found that the sugar-stimulated ethylene production could be observed also with intact leaves (unpublished results). We have therefore examined whether sucrose, when applied to cotton plants in the field, could enhance ethylene production, thereby stimulating their leaf defoliation. Accordingly, cotton plants, ready for harvest, were sprayed with commercial defoliants in combination with sucrose (Table 2). Ethylene production of the cotton leaves was measured 24 h after the field was sprayed. The defoliation index was evaluated one and two weeks after the treatments. The commercial defoliant $Mg(ClO_3)_2$ (Mag 40), increased ethylene production by 10-fold and addition of sucrose doubled its effect. One single spray of Mag 40 alone was insufficient to induce complete defoliation, but this could be achieved by spraying with a combination of Mag 40 plus sucrose or Ethrel. Although Ethrel by itself released high levels of ethylene, it seems that most of the generated ethylene was ineffective in inducing defoliation. On the other hand, sucrose was rather efficient in causing defoliation since it stimulated endogenous ethylene production, which was apparently more active in the abscission process.

TABLE 2. Stimulating effect of sucrose on rates of ethylene production and defoliation in cotton plants

Treatment	C_2H_4 Production 24h ($nl.g^{-1}.h^{-1}$)	Defoliation index[a] 7d	14d
Control	2	1.0	1.0
Mag 40[b]	25	2.0	3.5
Mag 40+sucrose 6%	50	4.5	5.0
Mag 40+Ethrel 0.4%	120	4.5	4.5
Ethrel 0.4%	127	2.0	2.0

[a] 1 = minimal defoliation; 5 = complete defoliation.

[b] $Mg(ClO_3)_2$; 1.3 liter/dunam.

4. CONCLUSIONS

The results reported in this paper imply that in many leaves IAA conjugates play a considerable role in the control of ethylene production. Carbohydrates could be involved in this control mechanism by enhancing the enzymatic hydrolysis of the IAA conjugates to yield free IAA, which, in turn, stimulates ethylene production on the step of ACC formation (13). The carbohydrates can stimulate also the final step of ethylene biosynthesis pathway, acting in synergism with CO_2. Since twelve naturally occurring carbohydrates were found to stimulate ethylene production in our studies (12), it is likely that there is no specificity for any of the

138

sugars tested. Thus, it is plausible that besides the above-mentioned mechanisms, the sugars can serve as energy suppliers, both in auxin metabolism and in the ethylene biosynthesis per se. Regarding the applicative use of sucrose in defoliation of cotton leaves, it can be further employed as a potential agent for inducing other ethylene-mediated phenomena in plant systems.

REFERENCES

1. Aharoni N, Anderson JM, Lieberman M. 1979. Production and action of ethylene in senescing leaf discs: effect of indoleacetic acid, kinetin, silver ion and carbon dioxide. Plant Physiol. 64:805-809.
2. Aharoni N, Lieberman M, Sisler HD. 1979. Patterns of ethylene production in senescing leaves. Plant Physiol. 64:796-800.
3. Aharoni N, Yang SF. 1983. Auxin-induced ethylene production as related to auxin metabolism in leaf discs of tobacco and sugar beet. Plant Physiol. 73:598-604.
4. Cohen JD, Bandurski RS. 1982. Chemistry and physiology of the bound auxins. Ann. Rev. Plant Physiol. 33:403-430.
5. Hangarter RP, Good NE. 1981. Evidence that IAA conjugates are slow-release sources of free IAA in plant tissues. Plant Physiol. 68: 1424-1427.
6. Hangarter RP, Peterson MD, Good NE. 1980. Biological activity of indoleacetylamino acids and their use as auxins in tissue culture. Plant Physiol. 65:761-767.
7. Kang BG, Newcomb W, Burg SP. 1971. Mechanism of auxin-induced ethylene production. Plant Physiol. 47:504-509.
8. Kao CH, Yang SF. 1982. Light inhibition of the conversion of 1-aminocyclopropane-1-carboxylic acid to ethylene in leaves is mediated through carbon dioxide. Planta 155:261-266.
9. Lau OL, Yang SF. 1973. Mechanism of a synergistic effect of kinetin on auxin-induced ethylene production. Plant Physiol. 51:1011-1014.
10. Lieberman M. 1979. Biosynthesis and action of ethylene. Ann. Rev. Plant Physiol. 30: 533-591.
11. Meir S, Philosoph-Hadas M, Aharoni N. 1983. Mode of action of sucrose in stimulating IAA-induced ethylene production by tobacco leaf discs. Plant Physiol. Suppl. 72:39.
12. Philosoph-Hadas S, Meir S, Aharoni N. 1983. Carbohydrates stimulate ethylene biosynthesis in tobacco leaf discs. Plant Physiol. Suppl. 72:121.
13. Yang SF, Adams DO, Lizada C, Yu YB, Bradford KJ, Cameron AC, Hoffman NE. 1980. Mechanism and regulation of ethylene biosynthesis. In, F. Skoog, ed. Proc. 10th Int'l. Conf. Plant Growth Substances, Springer Verlag, Berlin, pp. 219-229.

WOUND-INDUCED INCREASE IN 1-AMINOCYCLOPROPANE-1-CARBOXYLATE SYNTHASE
ACTIVITY: REGULATORY ASPECTS AND MEMBRANE ASSOCIATION OF THE ENZYME

AUTAR K. MATTOO[1,2] and JAMES D. ANDERSON[2]

Department of Botany, University of Maryland, College Park[1] and Plant
Hormone Laboratory[2], USDA, BARC(W), Beltsville, MD 20705, USA.

INTRODUCTION

The demonstration of 1-aminocyclopropane-1-carboxylic acid (ACC) as an
immediate precursor of ethylene biosynthesis in higher plants (1,2) and
of enzymatic conversion of S-adenosylmethionine (SAM) to ACC by cell-free
extracts of tomato fruit (3) established the following metabolic
sequence:

$$Methionine \longrightarrow SAM \longrightarrow ACC \longrightarrow ethylene$$

The conversion of SAM to ACC in vivo appears to be one of the main,
rate limiting steps in ethylene biosynthesis. This is apparent from
observations showing stimulation of ethylene production as being linked
to concomitant increase in the endogenous levels of ACC (4). Thus,
studies aimed at understanding the regulation of the ACC synthesizing
enzyme (ACC synthase) are of much current interest. The tissue which, on
wounding, consistently yields a markedly increased activity of ACC
synthase is tomato fruit (3-5). The mechanism that underlies this marked
increase in extractable ACC synthase activity is not understood, although
wounding seems to initiate the synthesis of new proteins and discrete
translatable mRNA's (Mattoo A.K., Anderson, J.D. Nakhasi, H.L, 1983,
unpublished data).

INVOLVEMENT OF MEMBRANE WITH ETHYLENE BIOSYNTHESIS

The cellular membrane is, presumably, one of the sites where external
stimuli influence ethylene production. Membrane perturbants or osmotic
shock of tissue segments have been shown to cause inhibition of ethylene
biosynthesis (6,8). Short-term incubation of apple slices with
deoxycholate or Triton X-100 in isotonic medium at 0°C or 22°C markedly
inhibits their subsequent ethylene production (Fig.1). Interestingly,
cirrasol, another detergent which by itself was without an effect,

Y. Fuchs and E. Chalutz (eds.) Ethylene: Biochemical, Physiological and Applied Aspects.
ISBN 90-247-2984-X. Printed in The Netherlands
©1984, Martinus Nijhoff/Dr W. Junk Publishers, The Hague.

partially prevented the deoxycholate-mediated inhibition of ethylene biosynthesis (Fig. 1).

Discontinuity in Arrhenius plots of incubation temperature against ethylene production was also suggested (9) as evidence of membrane involvement. The uncoupler, 2,4-dinitrophenol, which is known to inhibit ethylene biosynthesis (4, 10), changes the Arrhenius plot of ethylene production in apple but less so in pink tomato slices (Fig. 2). In both cases, this uncoupler at 10-20 µM is a potent inhibitor at all the temperatures tested. These data further support the contention that membrane function is involved in the regulation of ethylene production. It has been suggested that 2,4-dinitrophenol inhibits the conversion of ACC to ethylene (10).

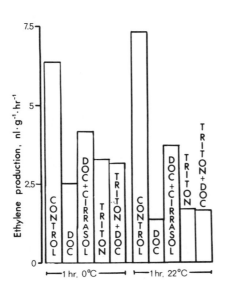

Fig.1. Inhibition by deoxycholate (DOC) and Triton X-100 of ethylene production by apple plugs. Apple plugs were incubated for 1hr at 0° or 22°C in the absence or presence of the indicated detergents. After each treatment, plugs were washed with medium without detergents, the containers sealed for 1 hr and ethylene production quantified by gas chromatography.

SOLUBLE AND MEMBRANE-ASSOCIATED ACC SYNTHASE

We have used the "wounding phenomenon" (11) to test various effectors of ethylene biosyhthesis as to their influence on the marked increase in ACC synthase activity seen when tomato (cv. Pikred) fruit slices are incubated in a solution of 600 mM sorbitol-10 mM Mes, pH 6.0 (Fig. 3). A survey carried out in this laboratory (Adams, Wang and Lieberman, 1980-81, unpublished) revealed that field-grown tomato varieties differ considerably in their content of extractable ACC synthase activity,

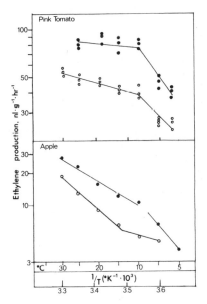

Fig. 2. Arrhenius plots of incubation temperature against rates of ethylene production by pink tomato and apple slices, and the effect of the uncoupler, 2,4-dinitrophenol. ●---●, Control; o---o, plus 2,4-dinitrophenol. The concentration of the uncoupler was 10 μM for apple slices and 20 μM for tomato slices.

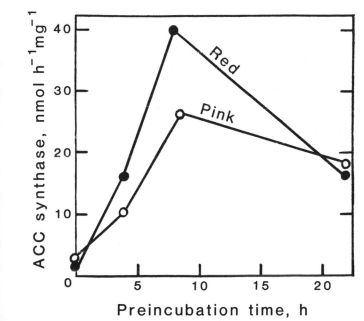

Fig.3. ACC synthase activity of tomato (cv.Pikred) fruit slices incubated for the times indicated at 23°C with 600 mM sorbitol-10 mM Mes, pH 6.0. After each incubation, the samples were frozen at -20°C, homogenized, and the clarified extracts gel-filtered and assayed for ACC synthase activity as previously described (5).

ranging from 17-32 nmol mg^{-1} h^{-1} in Duke, PSR 25237, Westover and Calypso to 100-200 nmol mg^{-1} h^{-1} in Pole King, Supersonic, Jet Sar, Pikred and Baron. For our studies, we routinely use greenhouse-grown Pikred variety (5), the ripe fruit of which, on wounding, yields extractable ACC synthase activity ranging from 30-50 nmol mg^{-1} h^{-1}. The time of wounding period, the ripening stage of the fruit, the conditions of plant growth and fruiting, and the method used for protein determination all are factors that influence the final specific activity of the enzyme recovered.

Fractionation of the homogenates prepared from wounded fruit showed that a part of the ACC synthase is membrane-associated. About 14% of the total enzyme activity from wounded green tomato was recovered in the 100,000 X g residue fraction and 68% in the soluble fraction. In wounded red tomato the recovered membrane-associated ACC synthase activity decreased to 2-3% and the soluble enzyme activity increased to 88% of the total.

The membrane-associated ACC synthase showed a higher affinity for SAM (apparent Km = 8 μM) than did the soluble enzyme (apparent Km = 18 μM). Isoelectric focusing of ripe tomato homogenate on a sucrose/ampholine column revealed two charge forms of the enzyme with pI's at 7.0 and 9.2. We have not determined which of these two forms represents the membrane-associated ACC synthase.

Spermine (12-15), carbonyl cyanide m-chlorophenylhydrazone (CCCP)(16) — a proton conductor that abolishes both electrical and pH gradients across membranes (17), and osmotic shock (6,18,19) were previously shown to inhibit ethylene production. As shown in Table 1, these effectors inhibit the wound-induced increase in ACC synthase activity, each to a different extent. Both the soluble and particulate ACC synthase activity were reduced to a similar extent (∿50%) when the tissue was osmotically shocked. However, spermine had a relatively more dramatic effect in reducing the particulate ACC synthase activity, while the effect of CCCP was more dramatic on the soluble form, presumably by reducing the ATP level required for its synthesis. Treatment with TBPB (tetrabutylammonium bromide), which is known to abolish electrical but not pH gradients across membranes (20), resulted in a more than 100% increase (over controls) in the particulate ACC synthase activity while causing about 50% decrease in the soluble form

Table 1. Differential effects of various treatments on the relative
distribution of particulate and soluble ACC synthase activity in tomato
fruit slices.

Treatment	ACC Synthase Activity (nmol h^{-1} mg^{-1} protein)	
	Particulate	Soluble
Wounded tissue, control	1.00	20.5
Plus spermine (10 mM)	0.19	9.8
Plus CCCP (50 µM)	0.39	2.5
Plus TBPB (10 mM)	2.30	9.3
Osmotic shock	0.52	11.7

(Table 1). These data highlight the dynamic control by membrane
potential on the induction and distribution of ACC synthase activity and
bring into focus the requirement for energy in causing this induction.
Also, the data on the effect of spermine confirm previous reports (15,21)
suggesting that the conversion of SAM to ACC is a site of spermine
inhibition.

PROTEASE INHIBITORS AND INDUCTION OF ACC SYNTHASE

The increase in the ACC synthase activity was inhibited by some
protease inhibitors, viz. phenylmethanesulfonyl fluoride (PMSF) and
soybean trypsin inhibitor, while other protease inhibitors, trasylol and
acetylaminobenzamide, actually further stimulated the rise (Table 2). We
have also found that PMSF inhibits ethylene biosynthesis in apple slices
(data not shown), the Cellulysin-mediated increases in ACC level and the
rate of ethylene production (22). The latter induction process is

Table 2. Effect of various inhibitors of proteases on the wound-induced
increase in soluble ACC synthase activity in tomato fruit slices.

Incubation Medium	ACC Synthase Activity	
	nmol h^{-1} mg^{-1} protein	% of control
Control (600 mM sorbitol 10 mM MES, pH 6.0)	4.97*	100.0
Plus PMSF (1 mM)	1.65	33.2
Plus trypsin inhibitor (5mg/ml)	2.51	50.5
Plus trasylol (200 kiu/ml)	9.35	188.0

*Overripe, post-climacteric, tomato fruit were used. ACC synthase
activity of the non-wounded tissue was 0.34 nmol h^{-1} mg^{-1} protein

also inhibited by soybean trypsin inhibitor (22). PMSF at 0.1 and 1.0 mM, trasylol at 20 and 200 kiu/ml, and soybean trypsin inhibitor did not inhibit the enzyme activity _in vitro_ or the chemical hydrolysis of ACC. The data suggest that specific proteolytic activity _in vivo_ is associated with the ethylene induction processes, possibly with the activation and/or inactivation of ACC synthase, or with the induction signal itself.

A MODEL FOR REGULATION OF ACC SYNTHASE AND ETHYLENE PRODUCTION

A model showing possible sites of control of the ethylene biosynthetic pathway is diagrammed in Fig. 4. Various stimuli external to the plant cell that affect the induction of ACC synthase include hormones such as IAA, stress such as wounding, and enzymes like Cellulysin (4,22,23). As shown here, the cellular membrane is possibly the target site where each stimulus is perceived since the _in situ_ increase in ACC synthase activity is inhibited by agents that: affect membrane potential (e.g. osmotic shock, CCCP), are impermeable (e.g. soybean trypsin inhibitor) or are highly charged (e.g. spermine). In this model, the result of cellular perception of these stimuli is a signal that controls _de novo_ synthesis of ACC synthase or activation of rapidly-turning-over 'inactive' enzyme. The evidence for _de novo_ synthesis during wounding is based on inhibitor data with cycloheximide (24,25) and density labeling with ^2H_2O (24). However, these latter data do not rule out the possibility of activation of a preexisting, rapidly-turning-over 'inactive' ACC synthase. The idea concerning activation of an inactive enzyme is based on the data showing inability of cycloheximide to inhibit Cellulysin-mediated induction of ACC synthesis (22), and inhibition by protease inhibitors of the latter phenomenon (22) and of ACC synthase induction in wounded tomato (this report). Nonetheless, it appears that proteolytic activity may be a prerequisite for induction of ACC synthase.

The newly synthesized or 'activated' ACC synthase may associate with membranes, a process stimulated when electrical gradients across membranes are abolished, e.g. by TBPB. In addition to the evidence presented here, certain characteristics of ACC synthase favour its association with cellular membranes. For example, ACC synthase is hydrophobic in nature (5, 24), is highly charged as is evident from a pI of 9.2 (this report), and its isolation from some tissues requires the presence of membrane-dissolving detergents in the isolation buffers (26). The membrane-

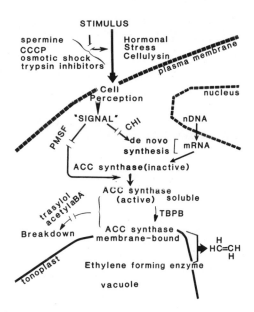

Fig. 4. A scheme illustrating various sites and levels of sub-cellular control of ethylene biosynthesis in higher plants.

associated ACC synthase may interact with the membrane-bound, ethylene forming enzyme. During this interaction, the product of the former enzyme, i.e. ACC, is synthesized in close proximity of the ethylene forming enzyme which uses ACC to produce ethylene. The higher affinity of membrane-associated ACC synthase for SAM could be advantageous in this regard. ACC has been shown to accumulate in the vacuole where ethylene forming enzyme also seems to reside (27). A close association of ACC synthase and the ethylene forming enzyme is supported by our data showing more ethylene production on Cellulysin induction than by feeding exogenous ACC (see Table 4 in ref. 28). We speculate that the membrane with which ACC synthase may more readily associate is the tonoplast.

The ACC synthase activity seems short lived (24). A recovery of twice as much ACC synthase from wounded tissue in the presence of trasylol as in the control may be explained by inhibition of breakdown of this enzyme. Future experimentation, isolation of the gene for ACC synthase and availability of antisera to the enzyme involved should help elucidate the validity of the model presented here.

146

ACKNOWLEDGMENTS

We thank Cathy Conlon and Annette Thomas for excellent technical assistance, Tom Barksdale for growing tomato plants, G. Bufler for making available his manuscript prior to publication, and Marvin Edelman and Janet Slovin for their comments on the manuscript. This investigation was supported by a grant from U.S.-Israel Binational Agricultural Research and Development Fund (BARD) and by Cooperative Agreement No. 58-32U4-1-216 between the USDA, Agricultural Research Service and the University of Maryland. Mention of a company name or trademark does not constitute endorsement by the US Department of Agriculture over other of a similar nature not mentioned.

LITERATURE CITED

1. Adams DO, Yang SF. 1979. Ethylene biosynthesis: identification of 1-aminocyclopropane-1-carboxylic acid as an intermediate in the conversion of methionine to ethylene. Proc. Natl. Acad. Sci., USA 76, 170-174.
2. Lurssen K, Naumann K, Schroder R. 1979. 1-Aminocyclopropane-1-carboxylic acid -- an intermediate of the ethylene biosynthesis in higher plants. Z. Pflanzenphysiol. 92: 285-294.
3. Boller T, Herner RC, Kende H. 1979. Assay for and enzymatic formation of an ethylene precursor, 1-aminocyclopropane-1-carboxylic acid. Planta 145: 293-303.
4. Yang SF. 1980. Regulation of ethylene biosynthesis. HortSci. 15: 238-243.
5. Mattoo, AK, Adams DO, Patterson GW, Lieberman M. 1982/83. Inhibition of 1-aminocyclopropane-1-carboxylic acid synthase by phenothiazines. Plant Sci. Lett. 28: 173-179.
6. Mattoo AK, Lieberman M. 1977. Localization of the ethylene synthesizing system in apple tissue. Plant Physiol. 60: 794-799.
7. Odawara S, Watanabe A, Imaseki H. 1977. Involvement of cellular membrane in regulation of ethylene production. Plant Cell Physiol. 18: 569-575.
8. Mattoo AK, Chalutz E, Lieberman M. 1979. Effect of lipophilic and water-soluble membrane probes on ethylene synthesis in apple and Penicillium digitatum . Plant Cell Physiol. 20: 1097-1106.
9. Mattoo AK, Baker JE, Chalutz E, Lieberman M. 1977. Effect of temperature on the ethylene-synthesizing systems in apple, tomato and Penicillium digitatum. Plant Cell Physiol. 18: 715-719.
10. Yu YB, Adams DO, Yang SF. 1980. Inhibition of ethylene production by 2,4-dinitrophenol and high temperature. Plant Physiol. 66: 286-290.
11. Boller T, Kende H. 1980. Regulation of wound ethylene synthesis in plants. Nature 286: 259-260.
12. Apelbaum A, Burgoon AC, Anderson, JD, Lieberman M, Ben-Arie R, Mattoo AK. 1981. Polyamines inhibit biosynthesis of ethylene in higher plant tissue and fruit protoplasts. Plant Physiol. 68: 453-456.

13. Suttle JC. 1981. Effect of polyamines on ethylene production. Phytochemistry 20, 1477-1480.
14. Ben-Arie, R, Lurie S, Mattoo AK. 1982. Temperature-dependent inhibitory effects of calcium and spermine on ethylene biosynthesis in apple discs correlate with changes in microsomal membrane viscosity. Plant Sci. Lett. 24: 239-247.
15. Even-Chen Z, Mattoo AK, Goren R. 1982. Inhibition of ethylene biosynthesis by aminoethoxyvinylglycine and by polyamines shunts label from 3,4-[^{14}C]methionine into spermidine in aged orange peel discs. Plant Physiol. 69: 385-388.
16. Apelbaum A, Wang SY, Burgoon AC, Baker JE, Lieberman M. 1981. Inhibition of the conversion of 1-aminocyclopropane-1-carboxylic acid to ethylene by structural analogs, inhibitors of electron transfer, uncouplers of oxidative phosphorylation, and free radical scavengers. Plant Physiol. 67: 74-79.
17. Sauer FD, Erfle JD, Mahadevan S. 1979. Methane synthesis without the addition of adenosine triphosphate by cell membranes isolated from Methanobacterium ruminantium. Biochem. J. 178:165-172.
18. Burg SP, Thimann KV. 1960. Studies on the ethylene production of apple tissue. Plant Physiol. 35: 24-35.
19. Imaseki H, Watanabe A. 1978. Inhibition of ethylene production by osmotic shock. Further evidence for membrane control of ethylene production. Plant Cell Physiol. 19: 345-348.
20. Grinius LL, Jasaitis AA, Kadziauskas YP, Liberman EA, Skulachev VP, Topali VP, Tsofiana LM, Vladimirova MA. 1970. Conversion of bio-membrane produced energy into electric form I. Submitochondrial particles. Biochim. Biophys. Acta 216: 1-12.
21. Fuhrer J, Kaur-Sawhney R, Shih L-M, Galston AW. 1982. Effects of exogenous 1,3-diaminopropane and spermidine on senescence of oat leaves. II. Inhibition of ethylene biosynthesis and possible mode of action. Plant Physiol. 70: 1597-1600.
22. Anderson JD, Mattoo AK, Lieberman M. 1982. Induction of ethylene biosynthesis in tobacco leaf discs by cell wall digesting enzymes. Biochem. Biophys. Res. Commun. 107: 588-596.
23. Lieberman M. 1979. Biosynthesis and action of ethylene. Annu. Rev. Plant Physiol. 30: 533-591.
24. Acaster MA, Kende H. 1983. Properties and partial purification of 1-aminocyclopropane-1-carboxylate synthase. Plant Physiol. 72: 139-145.
25. Hyodo H, Tanaka K, Watanabe A. 1983. Wound-induced ethylene production and 1-aminocyclopropane-1-carboxylic acid synthase in mesocarp tissue of winter squash fruit. Plant Cell Physiol. 24, 963-969.
26. Bufler G, Bangerth F. 1984. Effect of propylene and oxygen on the ethylene-producing system of apples. In press.
27. Guy M, Kende H. 1983. Ethylene formation in pea protoplasts. Plant Physiol. Supp. 72:209.
28. Anderson JD, Chalutz E, Mattoo AK. 1984. Purification and properties of the ethylene-inducing factor from the cell wall digesting mixture, Cellulysin. These proceedings, in press.

REDUCED S-ADENOSYLMETHIONINE DECARBOXYLASE ACTIVITY IN ETHYLENE TREATED
ETIOLATED PEA SEEDLINGS

AKIVA APELBAUM, ISAAC ICEKSON AND ARIE GOLDLUST
DEPT. OF FRUIT AND VEGETABLE STORAGE, AGRICULTURAL RESEARCH ORGANIZATION,
THE VOLCANI CENTER, P.O. BOX 6, BET DAGAN 50250, ISRAEL

INTRODUCTION

Polyamines are widely distributed throughout microbial, animal and
plant cells. They are universally associated with plant growth and cell
division (1,2) and recently the role of a "second messenger" has been
proposed for them (3). In microbial and plant cells, putrescine is
synthesized either from arginine via agmatine or directly from ornithine.
The putrescine formed serves as an acceptor for a propylamino moiety to
form spermidine, and spermidine in turn serves as an acceptor of a second
propylamino moiety to form spermine (4). The origin of the propylamino
transferred is S-adenosylmethylthiopropylamine, formed from S-adenosylmethionine (SAM), through the action of S-adenosylmethionine decarboxylase (E.C.4.1.1.50) (SAMDC) (5).

SAM appears to be a metabolite of central importance in plant
physiology being at the same time precursor of a) ACC and consequently
ethylene production (6); b) spermidine and spermine (7); and c) methylated
bases in DNA and RNA, like 5-methylcytosine in pea DNA (8).

Since ethylene has been associated with the arrest of cell division
and growth in pea seedlings (9) and polyamines are on the contrary
generally associated with areas where a high rate of cell division is
taking place (10,11), the fate of SAM and the direction in which the
dynamic equilibrium operating in the cell is turned, would be crucial in
determining whether cell division and growth, or arrest of growth and
senescence, would take place. These two directions seem to be closely

Abbreviations:

SAM, S-adenosylemethionine; SAMDC, S-adenosylmethionine decarboxylase;
ADC, arginine decarboxylase; ODC, ornithine decarboxylase; α-DFMA,
DL-α-difluoromethylarginine; α-DFMO, DL-α-difluoromethylornithine;
MGBG, methyl glyoxal-bis-guanyl hydrazone.

Y. Fuchs and E. Chalutz (eds.) Ethylene: Biochemical, Physiological and Applied Aspects.
ISBN 90-247-2984-X. Printed in The Netherlands
©1984, Martinus Nijhoff/Dr W. Junk Publishers, The Hague.

interrelated at points other than having SAM as a common precursor. It has been already shown that polyamines inhibit ethylene formation in fruits (12,13) and meristematic tissues (14). In this study we present results on the effect of ethylene on pea seedling, SAMDC activity, showing that ethylene provokes a pronounced and reversible inhibition of SAMDC activity.

MATERIALS AND METHODS

Plant material

Pea seeds (Pisum sativum var. 'Kelvedon Wonder') were soaked in water for 6 h and sown in moist vermiculite. Seedlings were grown in the dark at $22^{O}C$ with 80% R.H. for 3-6 days before use. Pots of seedlings also were grown under hypobaric conditions by placing them in a 10-liter desiccator which was evacuated continuously at approximately 0.5 standard cubic foot per h with a vented exhaust oil-seal pump. The pressure within the desiccator was maintained at 120 mm Hg by continuously admitting pure O_2 to the desiccator through a Matheson No. 49 regulator. The incoming O_2 was saturated at the reduced pressure passing it through water. For ethylene treatments, seedlings constantly kept in the dark, were placed in a 10 l desiccator connected to an inlet of a water-saturated mixture of air and ethylene (100 μl/l) with a flow rate of 0.1 l/min.

Extraction and assay of SAMDC activity. Excised plumular hooks were homogenized with five volumes of 0.05 M potassium phosphate buffer pH 7.4 containing 0.2 mM EDTA, in a ground glass homogenizer at $0-4^{O}C$. Homogenates were centrifuged at 39,000xg for 20 min and the supernatants were used for SAMDC activity determination.

The standard assay medium for SAM decarboxylation (15) contained the following ingredients in a total volume of 0.5 ml:60 μmoles sodium phosphate buffer pH 7.2, 0.1 μmole S-adenosylmethionine containing 0.06 μCi of S-adenosyl-L-methionine (carboxyl ^{14}C) (New England Nuclear, spec. activity 52.6 mCi/mmole), 2.5 μmoles of dithiothreitol and enzyme solution (0.2-1.6 mg protein) in 250 μl of potassium phosphate buffer (0.05 M, pH 7.4), containing 0.2 mM EDTA.

Incubations were performed at $37^{O}C$ for 30 minutes, in a tube sealed with a rubber cap fitted with a polypropylene center well, containing a paper wick impregnated with 0.25 ml Soluene 350 (Packard Inc.). Incubations were terminated by injection of 0.2 ml of 6N H_2SO_4, and tubes were further incubated for 60 min at the same temperature in order to release

all $^{14}CO_2$ from the reaction mixture. At the end of this period, the
center wells were transferred to a plastic scintillation vial containing
4.5 ml Aqualuma Plus scintillation liquid (Lumac B.V., Holland). The
vials were counted in a Kontron liquid scintillation spectrometer. In all
assays care was taken to work under conditions of linear response to the
protein concentration and with time. Blanks with acid-treated or pre-
boiled enzyme were currently run in each experiment and their values
substracted. Spontaneous decarboxylation of SAM in the absence of the
enzyme was checked periodically under the standard assay conditions, the
values obtained were the same as those from acid-treated or boiled enzyme
controls. Protein determinations were made in the 39,000xg supernatant
using Bradford's method (16). The data presented are from a single
experiment representative of a group of 3-4 experiments. Each measurement
was done in triplicate.

RESULTS

Characterization of the enzyme and specificity of the decarboxylation

SAMDC (S-adenosyl-L-methionine carboxylase E.C.4.1.1.50) activity can
be detected in extracts of meristematic tissues of pea seedlings. The
optimal conditions for the assay were found to be pH 7.4 and temperature
of 37^0C. The kinetic parameters of the reaction were studied and in the
crude extract SAMDC exhibited a typical Michaelis-Menten kinetics; the
apparent Km determined from a Lineweaver-Burke plot (Fig. 1) was 0.2 mM
and the apparent Vmax obtained was 4.76 nmoles/mg protein/h.

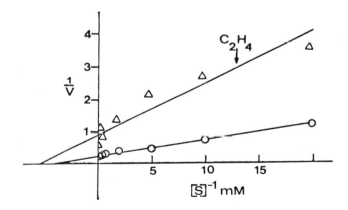

FIGURE 1. Lineweaver-Burke plot of SAMDC activity from control and
ethylene-treated pea seedlings.

Pea seedling SAMDC was not activated by either Mg^{++} ions or putrescine. Methyl glyoxal-bis-guanyl hydrazone (MGBG), a specific inhibitor of SAMDC (17), inhibited SAMDC from pea seedlings with an I_{50} of approximately 1 µM. Addition of ornithine or arginine or the inhibitors of ADC and ODC; DFMA or DFMO (1 mM) to the standard reaction mixture had no effect on SAMDC activity, thus ruling out a possible contribution of ADC or ODC activity to the decarboxylation of SAM.

Effect of ethylene on SAMDC activity. Exposure of pea seedlings to 50 µl/l ethylene resulted in inhibition of SAMDC activity. The time course of the effect of C_2H_4 on SAMDC activity is shown in Fig. 2, where a progressive inhibition of SAMDC activity is shown, reaching maximum effect (75% inhibition of the activity) within 18 h.

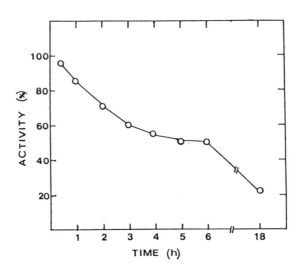

FIGURE 2. Time course of SAMDC inhibition by ethylene

Pea seedlings 3-4-days old were treated with ethylene (50 µl/l) for the indicated times. Control value is 3.2 nmoles/mg protein/h.

The kinetic parameters of the enzyme extracted from peas at the time of maximal inhibition by C_2H_4 are shown in Fig. 1. The apparent Km for the ethylene-treated enzyme was 0.18 mM as compared with 0.2 mM for the control preparations. The apparent Vmax of the ethylene-treated plants was markedly reduced; 1.8 nmoles/mg/h protein, as compared with the control enzyme; 4.76 nmoles/mg protein/h. Premixing of extracts from control and ethylene-treated plants always gave exactly additive results, suggesting

that the change in Vmax observed was not due to the appearance of a free
inhibitor or the loss of an activator molecule in the C_2H_4-treated seedlings.

There is a gradual recovery of the inhibition of SAMDC activity
produced by the ethylene treatment upon ventilation, as is shown in Fig. 3.
SAMDC activity reached pre-ethylene treatment values 5-6 h after ventilation indicating that seedlings are able to recover from the inhibition
produced by C_2H_4 treatment.

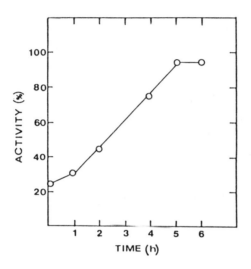

FIGURE 3.   Recovery of SAMDC activity from ethylene inhibition
     Plants (3-4-day old) treated with ethylene (100 μl/l) for 20 h.   They
were then ventilated for the indicated time.   Control value = 100% is the
value before inhibition by C_2H_4 (1.95 nmoles/mg protein/h).

The inhibition of SAMDC produced by ethylene is a concentration-
dependent phenomenon as can be seen in Fig. 4, approximately 50% inhibition
is obtained at a concentration of 0.2 μl/l, and maximal inhibition is
reached at 6 μl/l of C_2H_4; further increase in ethylene concentration to
100 μl/l did not have any additional effect.

154

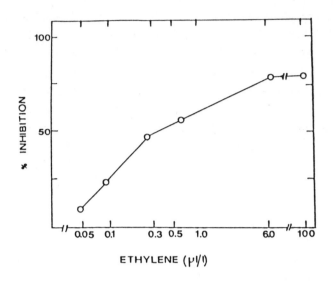

FIGURE 4. Ethylene inhibition of SAMDC: concentration curve.
       Pea seedlings 3-4-days old treated for 20 h with ethylene at the
indicated concentrations (control = 100%, 1.65 nmoles/mg protein/h).

       Treatments which cause a decrease in the endogenous ethylene levels
provoke an increase in SAMDC activity.
       When etiolated pea seedlings were grown for 24 h at subatmospheric
pressure (120 mmHg) or exposed for 20 h to light, SAMDC activity increased
by 30% and 48%, respectively.

       DISCUSSION
       Pea seedlings show the presence of a SAM decarboxylating activity that
has several characteristics similar to those described for other plant
SAMDC activity (15). SAMDC activity was measured by the determination of
$^{14}CO_2$ evolved from S-adenosylmethionine labelled in the carboxyl group.
Two sources of possible error have been pointed out (15) when measuring
SAMDC activity in plants by $^{14}CO_2$ evolution; a) non-enzymatic decarboxylation of SAM; b) non-specific enzymatic reactions. The possibility that
a significant part of the $^{14}CO_2$ evolved from carboxyl ^{14}C SAM was not
produced enzymatically was ruled out by the use of controls containing no
enzyme or containing boiled enzyme which gave negligible values. The
second possibility was ruled out by the use of MGBG, a specific inhibitor

of SAMDC.

The report by Suresh and Adiga on SAMDC activity of Lathyrus sativus seedlings (18) stresses the need of caution when interpreting results obtained with crude extracts of plants. They described a SAMDC activity that was putrescine-activated, but putrescine activation was due to a chemical decarboxylation of SAM by the H_2O_2 produced by diamine oxidase activity present in the extract using putrescine as substrate. These results prompted us to check the effect of Mg^{++} ions and putrescine on pea seedling SAMDC. Since our results showed that in this system SAMDC was not affected by either Mg^{++} or putrescine, this artifactual effect was ruled out. The same pattern has been observed in carrots, cabbage leaves (15) and corn (19).

Treatment of etiolated pea seedling resulted in inhibition of growth (9,24) and inhibition in SAMDC activity. The reduction in SAMDC specific activity caused by ethylene treatment always runs parallel to that in total activity, indicating that the modulation in the enzyme activity was not influenced by possible fluctuation in the total reserve protein contents of the plant during the treatment period.

It could also be shown that the change in SAMDC activity due to the hormonal treatment is not accompanied by the appearance of a free inhibitor, as revealed by the mixing experiments.

Plant growth regulators have been implicated in modulation of polyamine biosynthesis as suggested by: augmented polyamine content in IAA-stimulated Helianthus tuberosus explants in culture (20), GA-treated Lathirus sativus embryos (21) and cytokin-treated Cucumis sativus cotelydons in culture (22). In all these cases polyamine biosynthesis was enhanced by growth promoters. On the other hand, treatment with ABA, which inhibits growth and tissue expansion, inhibited ADC activity and amine content (22). This effect was completely reversed by cytokin, which is in line with the biological relationship between ABA and BA. This serves to emphasize the obligatory relationship between accelerated growth and enhanced polyamine biosynthesis. On the other hand, reduction in polyamine formation is associated with factors inflicting growth inhibition.

Ethylene has been known to suppress growth in etiolated pea seedlings and numerous other plants (23). The hormone has been shown to exert this effect by inhibiting cell division in the meristems of shoot apecies (9) and slowing down cell expansion in the subapical regions of these tissues (24). In shoot apecies of etiolated pea seedlings, a quantitative relationship was found (9) between the inhibition of DNA synthesis, cell division and growth caused by ethylene. Consequently, it was concluded that ethylene inhibits plant growth predominantly via inhibition of DNA synthesis. Since polyamines have been implicated to be required for DNA synthesis (3), it is proposed that ethylene inhibits DNA synthesis by blocking polyamine formation via inhibition of SAMDC activity. Indeed, SAMDC has a sensitivity to ethylene similar to that found for cell division and DNA synthesis.

From the fact that ethylene inhibits SAMDC activity within a relatively short time, that the concentrations required to exert this effect are similar to those found in the plant tissue, that recovery from the inhibition occurs shortly after removal of the hormone, and that, furthermore, the increase in SAMDC activity observed in response to treatments that cause reduction in the endogenous level of ethylene, it could be postulated that endogenous ethylene might regulate the activity of SAMDC in pea plant tissue.

REFERENCES

1. Cohen SS. 1971. Introduction to the Polyamines. Prentice-Hall Inc., Englewood Cliffs, NJ.
2. Bachrach U. 1973. Function of Naturally Occurring Polyamines. Academic Press, New York.
3. Galston AW. 1983. Polyamines as Modulators of Plant Development. BioScience 33:382-388.
4. Tabor H, Tabor CW. 1964. Spermidine, spermine and related amines. Pharac. Rev. 16:245-300.
5. Pegg AE, Williams-Ashman HG. 1969. On the role of S-adenosylmethionine in the biosynthesis of spermidine by rat prostate. J. biol. Chem. 244:682-693.
6. Yang SF. 1980. Regulation of ethylene biosynthesis. HortScience 15:238-243.
7. Baxter C, Coscia CJ. 1973. In vitro synthesis of spermidine in the higher plant Vinca rosea. Biochem. biophys. Res. Commun. 54:147-154.
8. Kalousek F, Morris NR. 1969. Deoxyribonucleic Acid Methylase Activity in Pea Seedlings. Science 164:721-722.
9. Apelbaum A, Burg SP. 1971. Effect of ethylene on cell division and deoxyribonucleic acid synthesis in Pisum sativum. Plant Physiol. 50:117-124.
10. Villanueva VR, Adlakha RC, Cantera-Soler AM. 1978. Changes in polyamine concentration during seed germination. Phytochemistry 17:1245-1249.
11. Bagni N, Serafini-Fracassini D. 1974. The Role of Polyamines as Growth Factors in Higher Plants and their Mechanism of Action. Proc. 8th Int. Conf. on Plant Growth Substances (1973). p.1205-1217. Hirokawa

Publishing Co., Tokyo.

12. Apelbaum A, Burgoon AC, Anderson JD, Lieberman M, Ben-Arie R, Mattoo AK. 1981. Polyamines inhibit biosynthesis of ethylene in higher plants. Plant Physiol. 68:453-456.

13. Apelbaum A, Icekson I, Burgoon AC, Lieberman M. 1982. Inhibition by Polyamines of Macromolecular Synthesis and its Implication for Ethylene Production and Senescence Processes. Plant Physiol. 70:1221-1223.

14. Suttle JC. 1981. Effect of polyamines on ethylene production. Phytochemistry 20:1477-1480.

15. Coppoc GL, Kallio P, Williams-Ashman HG. 1971. Characteristics of S-Adenosyl-L-Methionine decarboxylase from various organisms. Int. J. Biochem. 2:673-681.

16. Bradford MM. 1976. A rapid and sensitive method for quantitation of microgram quantities of protein using the principle of protein-dye binding. Anal. Biochem. 72:248-254.

17. Williams-Ashman HG, Schenone A. 1972. Methyl Glyoxal Bis (Guanyl Hydrazone) as a potent inhibitor of mammalian and yeast S-adenosyl-methionine decarboxylases. Biochim. biophys. Res. Commun. 46:288-295.

18. Suresh MR, Adiga PR. 1977. Putrescine-sensitive (Artifactual) and Insensitive (Biosynthetic) S-Adenosyl-L-Methionine Decarboxylase Activities of Lathyrus sativus Seedlings. Eur. J. Biochem. 79:511-518.

19. Suzuki Y, Hirasawa E. 1980. S-adenosylmethionine decarboxylase of corn seedlings. Plant Physiol. 66:1091-1094.

20. Bagni N, Fracassini DS. 1974. The role of polyamines as growth factors in higher plants and the mechanism of action. Plant growth substances. Hirokawa Publishing Company Inc., Tokyo.

21. Ramakrishna S. 1975. Ph.D. thesis. Indian Inst. Sci., Bangalore.

22. Suresh MR, Ramakrishna S, Adiga PR. 1978. Regulation of arginine decarboxylase and putrescine levels in Cucumis sativum cotyledons. Phytochem. 17:57-63.

23. Burg SP. 1973. Ethylene in plant growth. Proceedings of the National Academy of Science, USA. 70:591-597.

24. Apelbaum A, Burg SP. 1972. Effect of ethylene and 2-4-dichlorophenoxy-acetic acid on cellular expansion in Pisum sativum. Plant Physiol. 50:125-131.

1-AMINOCYCLOPROPANE-1-CARBOXYLIC ACID (ACC) AND ETHYLENE PRODUCTION
DURING SENESCENCE OF OAT LEAF SEGMENTS

R. PREGER AND S. GEPSTEIN, TECHNION - Israel Institute of Technology,
Haifa 32000, Israel.

INTRODUCTION

An increase which was followed by a decrease in endogenous ethylene
production was shown in oat leaf segments during the later stages of
senescence in light (Gepstein & Thimann, 1981); the rise in ethylene production occurred at the same time as chlorophyll degradation in leaves
started. It was also shown that addition of ethylene or ACC to leaf
segments hastened senescence while compounds which inhibited its production (AVG, Co^{2+}, Ag^+) delayed the process.

The main objective of this study was to find which step of ethylene
biosynthesis pathway in oat leaf segments is influenced by age.

RESULTS

FIGURE 1. Endogenous
ethylene evolution
($\bullet$), ethylene production from
exogenously applied
1 mM ACC ($\blacksquare$) and
ACC level ($\blacktriangle$)
during oat leaf
senescence in the
dark.

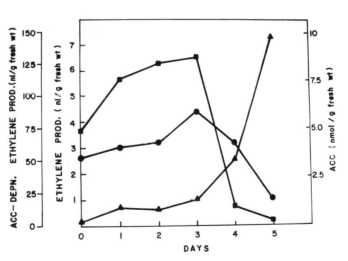

Y. Fuchs and E. Chalutz (eds.) Ethylene: Biochemical, Physiological and Applied Aspects.
ISBN 90-247-2984-X. Printed in The Netherlands
©1984, Martinus Nijhoff/Dr W. Junk Publishers, The Hague.

FIGURE 2. Changes in total
ACC (conjugated + unconjugated) (●) and in free ACC
(unconjugated) (o), during
oat leaf senescence in the
dark (———) and in the incubation medium of the leaves
(- - -).
The difference between total
and free ACC expresses the
amount of conjugated ACC
(malonyl-ACC).

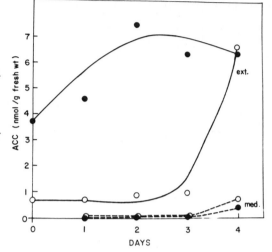

CONCLUSIONS AND SUMMARY

a. Endogenous and ACC-
dependent ethylene production during oat leaf
senescence show a climatic-
like pattern.

b. No correlation
between the decrease in ethylene formation and the high amount of ACC
was found in oat leaves during the latest stage of senescence.

c. There is no increase in ACC in the leaves and in the incubation
medium late in senescence.

The phenomenon of low ethylene production and high ACC content has
been demonstrated also in ripening fruits (Hoffman & Yang, 1980) and
during senescence of cut carnation flowers (Bufler et al., 1980). It may
be concluded that the conversion of ACC to ethylene rather than the
synthesis of ACC is the rate-limiting step during the later stages of
senescence in plants.

The drop in ethylene production late in senescence cannot be
accounted for either by a decrease in the amount of ACC or by a
conjugation of ACC. It may be suggested that the system which is
responsible for the conversion of ACC to ethylene deteriorates during
senescence.

REFERENCES

1. Bufler G, Mor Y, Reid MS, Yang SF. 1980. Planta 150: 439-442.
2. Gepstein S, Thimann KV. 1981. Planta 149: 196-199.
3. Hoffman NE, Yang SF. 1980. J. Am Soc. Hort. Sci. 105: 492-495.

EFFECTS OF NATURALLY PRODUCED ETHYLENE IN TISSUE CULTURE JARS.

Roberto JONA, Ivana GRIBAUDO, Rosanna VIGLIOCCO
Ist. Coltiv. Arboree - Università   Publication n. 556
Centro Miglioramento Genetico Vite - C.N.R. Publication n. 129
Via Pietro Giuria, 15,I - 10126   TORINO TO Italy

In tissue culture methods reported by literature, the size of the vessel and the kind of stopper are hardly mentioned. This affects the composition of the atmosphere inside the vessel. Ethylene could be a major factor among inner atmosphere components.

MATERIALS AND METHODS: a series of different vessels, containing the same medium, but stoppered in different ways were used: 1)160 ml jar, stoppered with a cotton plug.2)100 ml jar stoppered with a cotton plug.3)160 ml jar stoppered withouth gasket and wrapped with plastic film (fit foodstuffs).4)100 ml jar stoppered with air tight cover. Glass tubes stoppered with 5) cotton plug 6) metal cap (Cap-o-test), lined with cotton plug.7)Cap-o-test, wrapped with plastic film.8)Cap-o-test sealed with Parafilm.

The vegetal materials were 2 cm long shoots of GF677 peach x almond hybrids and leaf explants of Ficus lyrata,Ward. The media employed were:a) proliferating medium for GF677,b) rooting medium for GF677,c) proliferating medium for Ficus lyrata.

A new experiment used air tight vessels only: into half of the vessels an open vial was added before autoclaving the medium and about 2 gms of silica-$KMnO_4$ gel were poured into the open vial when shoots were implanted, in order to absorb ethylene. The metal cover of the jars was pierced and the hole was stoppered with a rubber stopper which, in turn, could be easily pierced by the needle of a syringe. Two weeks after implanting, a sample of 5 ml of the inner atmosphere was taken and injected into a gas chromatograph for ethylene measurements.

Y. Fuchs and E. Chalutz (eds.) Ethylene: Biochemical, Physiological and Applied Aspects.
ISBN 90-247-2984-X. Printed in The Netherlands
©1984, Martinus Nijhoff/Dr W. Junk Publishers, The Hague.

RESULTS: <u>GF677 Proliferation Medium</u>. In every vessel which was airtight a remarkable level of leaf epinasty followed by shriveling and eventually leaf drop was noticed. With a cotton plug, the leaves were initially larger and apparently more healthy. After a while however the medium was drying up, thus hindering further growth of the plantlets.<u>Rooting medium</u>.Limited closure with cover without gasket, wrapped whith plastic film appears to be the more suitable cover for rooting.

<u>Ficus lyrata</u>: air tight cover induces only proliferation of a whitish and spongious callus, while covers without gasket and wrapped with plastic film favour shoot proliferation in all explants. <u>Removal of ethylene by KMnO$_4$</u>: Proliferation rate of GF677 on BAP medium was much higher in vessels without KMnO$_4$. On the opposite, rooting of GF677 on IBA medium was almost three times with KMnO$_4$ as compared to vessels without KMnO$_4$.When this was absent a remarkable epinasty and shriveling were clearly noticeable.Gas chromatographic analysis showed that KMnO$_4$ reduces ethylene to a very low lovel. It should be noticed that within the same lag of time (l5 days) ethylene developed inside the jar is clearly dependent on the nature of the medium: much more where BAP is present as compared to IBA.

DISCUSSION. From these results it clearly appears that ethylene has both beneficial and negative effects on plant proliferation and development. Addition of KMnO$_4$ does not seem feasible in routine work. However a limited amount of ethylene may be useful for proliferation without inducing negative side effect.

As practical conclusion it can be argued that renewal of the medium may be important in order to remove the explants from an ethylene rich atmosphere. Furthermore the stopper and its gaseous permeability could be crucial for the success of the method. Flame sterilization could also affect the level of ethylene, because the latter is contained in the Bunsen's flame.

INVOLVEMENT OF ETHYLENE IN LIATRIS CORM DORMANCY.

Vered Keren – Paz and Amihud Borochov
Department of Ornamental Horticulture, Faculty of Agriculture, The Hebrew
University of Jerusalem, Rehovot, Israel.

INTRODUCTION

Liatris (<u>L. spicata</u>, cv. Callilepis) is a corm plant, grown for cut
flower production. The corms exhibit seasonal changes in dormancy as
expressed by the number and rate of shoot sprouting. Cold (2–4°C) storage
for several weeks is needed to break dormancy. It is widely accepted that
the control of dormancy in geophytes is mediated by hormones, particulary
by ethylene. The aim of this work was to study the possible relation
between ethylene and dormancy in liatris corm.

MATERIALS AND METHODS

Liatris corms were obtained from a commercial grower. ACC content
was measured according to Lizada and Yang (1) in the lyophilised tissue
extract. Ethylene forming enzyme activity of the microsomal membrane
fraction was measured according to Mayak et al. (2).

RESULTS AND DISCUSSION.

Dormant corms, held in cold storage for different periods of time,
showed a peak in ethylene production at 30°C at the time when dormancy
break was accomplished. Exogenously applied ethylene, given as a short
(1/2h) dip in ethrel solution, caused an increase in ethylene production
and corm sprouting. Both of these effects were proportional to the ethrel
concentration (0 – 4000 ppm), and independent of the original depth of
dormancy in the treated corms. The ethrel-induced ethylene production
declined gradually, 50% after 3 days, and was still evident after 7 days.

Isolated buds were much more active in producing ethylene as
compared with parenchyma disks of the same corms, both in high ethylene

Y. Fuchs and E. Chalutz (eds.) Ethylene: Biochemical, Physiological and Applied Aspects.
ISBN 90-247-2984-X. Printed in The Netherlands
©1984, Martinus Nijhoff/Dr W. Junk Publishers, The Hague.

producing corms and low ethylene producing corms. Incubation with ACC
solution resulted in higher ethylene production in both populations of
corms. However, even with ACC supply, buds were more actively producing
ethylene as compared with parenchyma. Endogenous ACC content of buds was
not correlated with their natural ethylene production. This led us to
evaluate the involvement of changes in ACC conversion to ethylene. A
microsomal membrane fraction, prepared from liatris corm buds, exhibited
ethylene forming enzyme activity when incubated with ACC.
Characterization of this activity revealed apparent similarity with that
of the reported ethylene forming enzyme of carnation petals (2).
Membrane ethylene forming enzyme activity of corm buds of different
natural ethylene production differed: ethylene forming enzyme activity of
high ethylene producing corms was ca. 3 times higher than that of low
ethylene producing corms.

In contrast to the exogenous ethylene effect, treatments with
inhibitors of ethylene synthesis (cobalt) or action (STS) inhibited
sprouting. In addition, cobalt inhibited ethylene production.
Cobalt and STS effects were expressed by decreasing both sprouting and
height.

We conclude that ethylene production is involved in the control of
the rate of growth of liatris buds (i.e. dormancy break), with the
limiting step of ethylene production being the activity of ethylene
forming enzyme.

REFERENCES.
1. Lizada M.C. and S.F.Yang (1979). A simple and sensitive assay for
1-aminocyclopropane-1-carboxylic acid. Anal. Biochem. 100:140-145.
2. Mayak S., R.L.Legge and J.E.Thompson (1981). Ethylene formation from
1-aminocyclopropane-1-carboxylic acid by microsomal membrane from
senescing carnation flowers. Planta 153:49-55.

ENDOGENOUS ETHYLENE PRODUCTION AND FLOWERING OF BROMELIACEAE.

M. DE PROFT, L. JACOBS and J.A. DE GREEF

Department of Biology, University Instelling Antwerpen, Universiteitsplein
1, B-2610, Wilrijk, Belgium

INTRODUCTION
For many years it is known that exogenous ethylene can induce flowering
of pineapple. Due to the lack of sensitive methods the ethylene production
of bromeliad plants has never been studied in relation to flowering induction.
We demonstrated a positive correlation between ethylene production capacity
of the bromeliad plant and its flowering process.

MATERIAL and METHODS
Plant material
One year old plants of Guzmania lingulata var. minor were used in
our experiments. They had a fresh weight of 25-35 grams. In Belgium these
plants are commercially available on large scale.
Ethylene measurements
Ethylene determinations were performed in an open flow system as
described earlier (1, 2). The plants were kept in a photoperiod of 12 hours.

RESULTS
The plants were treated with AVG, IAA, ACC or ethephon. The first three
chemicals were applied by pouring a 10 ml solution of 0.1 mM into the
calyx of the plants. The maximal ethylene release and the opening of the
first flower were then measured in function of time (Table 1). The highest
ethylene production obtained within a period of 30 days is given. The time
of opening of the first flower can be used as a measure of the effectiveness
of the treatment. After AVG treatment the ethylene production was reduced
by 50%. The auxin stimulated ethylene production was also abolished by
AVG pretreatment. Exogenous ACC increased the ethylene production
dramatically. The high degree of spontaneous flowering was totally
suppressed by AVG. Auxin stimulated flowering partially but either ACC

Y. Fuchs and E. Chalutz (eds.) Ethylene: Biochemical, Physiological and Applied Aspects.
ISBN 90-247-2984-X. Printed in The Netherlands
© 1984, Martinus Nijhoff/Dr W. Junk Publishers, The Hague.

166

or ethephon gave maximal flowering at the given concentrations.

Table 1 : Endogenous ethylene production and flowering induction of
Guzmania plants after AVG, IAA, ACC and ethephon treatments.

treatment	ethylene (nl/h.plant)	flowering (%)	first flower (days)
control	3.0±0.1	60	78±21
AVG 0.1mM	1.5±0.1	0	∞
IAA 0.1mM	12.9±0.5	80	75±20
ACC 0.1mM	55.3±0.5	-	-
ACC 0.4mM	-	100	75±13
AVG 0.1mM + IAA 0.1mM	1.5±0.1	-	-
AVG 0.1mM + ACC 0.4mM	290.0±0.5	100	76± 9
ethephon 500ppm	-	100	74± 9

CONCLUSION

Endogenous ethylene production capacity is positively correlated
with the flowering induction process of Guzmania lingulata var. minor.

ACKNOWLEDGEMENTS

This work is supported by the Belgian IWONL, grant n° 3620A.

AVG was a generous gift of Dr. A. Stempel, Hoffman-La-Roche, USA.

REFERENCES
1. De Greef J., De Proft M., De Winter F. 1976. Gaschromatographic
determination of ethylene in large air volumes at the fractional
parts-per-billion level. Anal. Chem. 48(1): 38-41.
2. De Greef J., De Proft M. 1978. Kinetic measurements of small ethylene
changes in an open system designed for plant physiological studies.
Physiol. Plant. 42: 79-84.

ACC : 1-aminocyclopropane-1-carboxylic acid
AVG : aminoethoxyvinyl glycine
IAA : indole acetic acid

ROLE OF ETHYLENE IN DISTRIBUTION OF ASSIMILATES IN CARNATIONS

H.VEEN and A.A.M.KWAKKENBOS

Center for Agrobiological Research, Bronsesteeg 65, P.O.B. 14,
6700 AA Wageningen, The Netherlands

INTRODUCTION

The present report deals with the question,if ethylene interferes
with the translocation of endogenously formed assimilation products.
Sink activity of petals and pistil was studied in time course
experiments as related to the source activity of the calyx.
Pre-treatment of flowers with silver thiosulphate (STS) caused
complete blocking of the action of ethylene (1).The ethylene-sensitive
components in source-sink relationships in the carnation flower can
thus be distinguished from the non-sensitive components.

PROCEDURE

Material. Carnations (Dianthus caryophyllus L.cv.White Sim)
were harvested in a nursery at the usual commercial stage of flowering.
Experiments were carried out in a climate-room.
2.2. Methods. Radioactive CO_2 was applied 1, 4, 7, and 11 days after
harvest to the whole cut flower. After an assimilation period of
30 min , 4 flowers were dissected;calyces,petals and pistils were
isolated. An other group of 4 flowers,also treated with $^{14}CO_2$, were
kept for 24 hours on distilled water and then dissected. All samples
were freeze-dried,powdered,weighed and oxidized in a Liquid Scintillation Oxidizer System. Radioactivity was counted in a Liquid Scintillation Counter.

RESULTS AND DISCUSSION

Calyces of flowers dissected after a labelling period of 30 min
showed a maximum fixation capacity at or before 4 days after harvest.
Differences between controls and STS pre-treated flowers were not
significant. Apparently, ethylene did not interfere in ^{14}C fixation

Y. Fuchs and E. Chalutz (eds.) Ethylene: Biochemical, Physiological and Applied Aspects.
ISBN 90-247-2984-X. Printed in The Netherlands
© 1984, Martinus Nijhoff/Dr W. Junk Publishers, The Hague.

in calyx tissue. In flowers dissected 24 hours after a 30 min labelling period, 20 per cent of the original fixed ^{14}C was recovered from the calyx. Simultaneously with petal wilting (after 6 days), this recovery percentage increased up to 80-90 per cent in controls. This increase in residual radioactivity in the calyx is probably caused by a loss of sink strength of the petal tissue. In STS pre-treated flowers, this percentage remained on a constant level: 20 per cent.

In petals from flowers immediately dissected after labelling, only small amounts of ^{14}C were recovered. Data from flowers, which were dissected 24 hours after labelling, showed that sink activity increased in petals. Simultaneously with wilting, this activity dropped to zero. The STS pre-treated flowers kept on attracting radioactivity and finally lost some of their sink strength after more than 12 days.

Pistils of flowers dissected immediately after labelling, showed negligible radioactivity. From data on flowers which were dissected 24 hours after labelling, it can be concluded that pistils from control flowers showed a constant sink activity in the first 7 days after harvest. Then, after wilting occurred, the sink strength increased. In flowers pre-treated with STS, pistils accumulated small quantities of ^{14}C within 24 hours after applying radioactive CO_2 to the flower. This took place only within the first 5 days after harvest. Thereafter the pistils of the STS pre-treated flowers completely lost their sink activity .

Our data on the sink activity of petals and pistils fit in with the conclusion of Cook and Van Staden (2) that "after petal wilting, the sink strength appeared to move from the petals to the ovary where the majority of the ^{14}C then accumulated. With STS the petals remain the active sink and ovary development is suppressed".

By regulating petal wilting, ethylene plays an important role in carbohydrate distribution within the flower.

A full report of this work will be published elsewhere .

REFERENCES

1. Veen H. 1979. Effects of silver on ethylene synthesis and action in cut carnations. Planta 145, 467-470.
2. Cook EL, Van Staden J. 1982/83 . Senescence of cut carnations flowers : ovary development and CO_2 fixation. Plant Growth Regulation 1 , 221-232 .

CHARACTERIZATION OF AN ENDOGENOUS INHIBITOR OF ETHYLENE BIOSYNTHESIS IN CARNATION PETALS.

Hanan Itzhaki, Amihud Borochov and Shimon Mayak, Department of Ornamental Horticulture, Faculty of Agriculture, The Hebrew University of Jerusalem, Rehovot, Israel.

INTRODUCTION

A membrane-bound enzymatic conversion of 1 – aminocyclopropane – 1 – carboxylic acid (ACC) to ethylene was recently demonstrated in carnation petals by Mayak et al. (1). They also reported the presence of a cytoplasmic inhibitor of this reaction. The aim of this work was to characterize this inhibitor.

MATERIALS AND METHODS

Enzyme and inhibitor preparation. Carnation (Dianthus caryophyllus, cv White Sim) petals were homogenated, centrifuged and fractionated as described before (1). Microsomal membranes were used as a source of enzymatic activity, and the combined supernatant as an inhibitor fraction.

Reaction system. The standard cell free reaction system was composed of 100 ug membrane protein, ACC at 4 mM, 5 ug MnCl and 50 mM Epps buffer at pH 8.5. Reactions were conducted for 1h at 30°C, and the ethylene produced was measured by gas chromatography.

RESULTS AND DISCUSSION

When a standard cell-free reaction system which included microsomal membranes from carnation petals as a source of enzymatic activity, and ACC as substrate was used, ethylene production was inhibited by addition of various amounts of cytoplasmic extract from the petals. For membranes, this inhibition was exponentialy dependent on the inhibitor amount, so 50% inhibition was achieved in the presence of the equivalent of 0.4 mg F.W. Similar effect on ethylene formation was found when petals were treated with inhibitor solution before being fed with ACC. In addition, ACC conversion to ethylene by both detergent – solubilized and intact membranes was inhibited to a similar extent by including cytoplasmic extract in the reaction mixture. The inhibition was

Y. Fuchs and E. Chalutz (eds.) Ethylene: Biochemical, Physiological and Applied Aspects.
ISBN 90-247-2984-X. Printed in The Netherlands
©1984, Martinus Nijhoff/Dr W. Junk Publishers, The Hague.

reversible, since high-speed centrifugation separated the inhibitor from the membranes. Through measuring enzyme activity at combinations of two substrate and four inhibitor concentrations we have created a Dixon plot, from which we were able to learn that the inhibitor is competitive, with Ki=eq. 0.8 mg F.W. The inhibitor does not change the optimal pH of the reaction, which was 9.0, but the inhibition was highly affected: While maximal inhibition was found at pH 8.0, a sharp decline was found at higher or lower pH velues. However, the inhibitory activity was not changed by short exposure to extreme pH values. The differences in inhibition at different pH values suggest change in inhibitor charge. Indeed, while the inhibitor was not adsorbed to a cation exchanger (CM-sephdex), it was strongly bound to an anion exchanger (DEAE-sephadex), and the binding was pH dependent. An isoelectric point of pH 3.2 was found. The inhibitor is dialysable. Based on gel filtration experiments, the M.W of the inhibitor was estimated as less than 1500, since it was excluded from the gel which has a characteristic separation limit of 1500 M.W. In addition, it was completly separated from proteins. The inhibitor is heat stable, and was not precipitated by TCA. Thus supporting a non-proteinaceous nature. The association of the inhibitor with material which absorbed at 280 nm, and concentration-dependent binding to PVP raised the possibility that the inhibition activity is at least in part due to a phenolic compound.

In summary, we interpet the above results as describing a natural endogenous inhibitor of ethylene synthesis which is a relatively small water-soluble anion at physiological pH, and which might include a phenolic ring. The inhibitor acts reversibly and it is competitive with ACC.

REFERENCES
1. Mayak et al. (1981).Ethylene formation from ACC by microsomes from carnation flowers. Planta 153:49-55.

REGULATION OF PATHOGENESIS AND SYMPTOM EXPRESSION IN DISEASED PLANTS
BY ETHYLENE

L.C. VAN LOON

Department of Plant Physiology, The Agricultural University,
6703 BD Wageningen, The Netherlands

An enhanced production of ethylene is associated with the development
of symptoms in plants infected with fungi, bacteria, viruses and viroids.
Symptoms such as inhibition of growth, epinasty, abscission of leaves and
organs, and chlorosis and premature senescence are part of the physiological responses of plants to various types of stress; these are similarly accompanied by stimulated ethylene production, suggesting that such
symptoms in diseased plants arise as a result of the stress imposed by
infection. This has led to the general conclusion that the evolution of
ethylene from diseased tissues most probably is of host origin arising as
a result of cellular damage. However, some pathogenic fungi and bacteria
are themselves capable of producing ethylene in vivo. Such ethylene may
act as a phytotoxin or modify the plant's response, enabling the pathogen
to overcome host defense mechanisms. In tulip bulbs infected with Fusari-
um oxysporum f. sp. tulipae the fungus produces high levels of ethylene,
resulting in storage diseases and shoot stunting and bud blasting upon
planting. In the absence of ethylene, a fungitoxic compound, tulipalin A,
is formed in the bulb scales after lifting. However, ethylene inhibits
the synthesis of tuliposide, the precursor of tulipalin, permitting the
fungus to penetrate the bulb (22). It has not been established, however,
whether the ethylene produced by Fusarium-infected tulip bulbs is derived
entirely from the fungus, or is, in part, synthesized by the bulb itself
in response to the ethylene produced by the pathogen.

The extremely limited amount of genetic information carried by viruses precludes them from producing ethylene and thus, in virus-infected
plants ethylene is produced solely as a plant reaction. Apart from being
associated with the types of symptom mentioned above, ethylene production
is increased in virus-infected plants notably when yellow chlorotic or
necrotic symptoms arise. Thus, the development of a light green - dark

Y. Fuchs and E. Chalutz (eds.) Ethylene: Biochemical, Physiological and Applied Aspects.
ISBN 90-247-2984-X. Printed in The Netherlands
© 1984, Martinus Nijhoff/Dr W. Junk Publishers, The Hague.

green mosaic on young leaves of tobacco infected with tobacco mosaic virus
(TMV) is not accompanied by increased ethylene production (15), whereas
ethylene is stimulated concomitant with the appearance of yellow chlorotic lesions on leaves of Tetragonia expansa due to infection with bean
yellow mosaic virus (BYMV) (18). The strongest stimulations occur when
necrosis develops, such as in tobacco cultivars carrying the gene N
after infection with TMV: a hypersensitive reaction ensues in which
necrotic local lesions are produced to which the virus remains confined.
It has been claimed that the ethylene produced during such a hypersensitive reaction of cucumber to cowpea mosaic virus is derived from linolenic acid through the action of lipoxygenase, activated by the cellular
damage accompanying lesion formation (23). However, during necrotic
local lesion formation in TMV-infected tobacco, no conversion of ^{14}C-
acetate, used to label fatty acids, into ethylene was observed (16).
Ethylene was produced as a sharp peak near the time of lesion appearance.
About 12 h before lesions became visible, the activity of the enzyme
1-aminocyclopropane-1-carboxylic acid (ACC)-synthase was increased up to
50-fold. As a result, ACC accumulated, until around the time of lesion
appearance, the activity of the enzyme system converting ACC into ethylene increased and release of ethylene occurred (14). The production of
virus-stimulated ethylene was blocked for 95% by aminoethoxyvinylglycine
(AVG). Moreover, when infected leaves were supplied with U-^{14}C-L-methi-
onine, the specific activity of the ethylene produced was in accordance
with that of the methionine pool, indicating that all the ethylene produced in hypersensitively reacting tobacco is derived from methionine
according to the same biosynthetic pathway as occurs in normal plant
tissues (16).

INVOLVEMENT OF ETHYLENE IN SYMPTOM DEVELOPMENT

Just as exogenously applied ethylene can cause senescence and abscission, it may induce some of the symptoms of disease in healthy plants,
suggesting that ethylene is involved in the regulation of pathogenesis
and symptom expression. Infection of cucumber seedlings by cucumber
mosaic virus (CMV) causes retardation of hypocotyl elongation. This was
found to correspond with an enhanced production of ethylene in infected
hypocotyls. Exogenously applied ethylene likewise retarded hypocotyl
elongation, whereas reduction of atmospheric ethylene with $KMnO_4$ to some

extent increased elongation, suggestive of a causative role of ethylene
in growth retardation (27). The ethylene produced was similarly implicated in the epinasty of (24) and development of chlorotic lesions on
CMV-infected cucumber cotyledons (26). Although lower levels of ethylene
were detected in the ambient atmosphere of diseased cotyledons, levels
of endogenous ethylene were significantly enhanced due to increased leaf
resistance to gaseous diffusion. Such effects seem to be confined to
fleshy organs, however, and are not apparent when ethylene emanation
from diseased leaves is examined.

When T. expansa plants inoculated with BYMV were exposed to 1% CO_2,
they failed to develop chlorotic lesions and symptomless infection en-
sued. However, when plants showing chlorotic lesions were gassed with
ethylene or sprayed with ethephon, necrotic spots developed within the
chlorotic lesions (4). Necrotic lesion production in TMV-infected Nicotiana glutinosa was inhibited by CO_2, and in plants of Havana 425 tobacco upon exposure of the inoculated leaves to 1% CO_2 the developing lesions were yellow instead of the usual brown (31). Such observations suggest that formation of chlorotic or necrotic lesions might represent
only a quantitative difference. Although treating tobacco leaves with
300 ppm ethylene for 40 h induced no necrosis, only moderate necrosis
(30), this does not rule out a role of ethylene in the necrotization
reaction. Ethylene production reaches a peak coincident with lesion formation, but it is already significantly enhanced 12 h earlier (14, 16,
18) and by that time, ACC-synthase activity is reaching a maximum. Ethylene production is similarly enhanced 8 h after inoculation of cowpea
leaves with CMV, 10 h before necrotic lesions become apparent (23). Thus,
the enhanced ethylene production is not the result of a wounding reaction
due to cellular necrotization. Moreover, in hypersensitively reacting
tobacco, the changes in ACC-synthase activity and ACC content are confined to the cells immediately adjacent to the point of entry of the
virus (15). Because these tissues comprise only a small part of the leaf,
probably not more than at most 1%, the actual increases in ACC-synthase
activity and ethylene production on a cell basis are at least two orders
of magnitude higher than measured over the whole leaf, representing local
increases of a few thousand-fold over water-inoculated control tissues.
The resulting high local concentrations of ethylene might lead to changes in membrane permeability and, thereby, initiate reactions ultimately

leading to visible tissue necrosis (45).

Treatment with ethylene may also aggrevate or reduce symptom expression in infected plants depending on the host - pathogen combination, the physiological state of the plant tissue and the time of application with respect to infection. As mentioned above, ethylene may block the synthesis of the antifungal tulipalin in tulip bulbs. Similarly, the softening of fruits ripening under the influence of ethylene may make them vulnerable to fungal or bacterial rots, whereas green fruits are not attacked (36). On the other hand, local treatment of N-gene-containing tobacco's with ethylene or ethephon induces an increased capacity to counteract the spread of TMV upon subsequent inoculation in all leaves present, resulting in the production of substantially smaller, and often also less, lesions (35, 44). This so-called systemic acquired resistance (34) is likewise induced by the virus itself -which induces large amounts of ethylene during primary lesion formation- and is more prominent in younger than in older leaves on the plant. The extent of systemic acquired resistance is correlated with an increased capacity to convert ACC to ethylene, ensuring a more rapid production of ethylene upon challenge inoculation (15). However, once lesions have become apparent and ethylene production is stimulated, exogenously applied ethylene may enhance further lesion growth (43). Since ethylene also promotes lesion expansion when leaves are detached, this effect appears to be due to ethylene-induced senescence around the developing lesions (18). Thus, the addition of ethylene prior to lesion appearance may stimulate the resistance mechanism, but it is unable to do so in infected plant parts once metabolism has been changed concomitant with lesion development.

ETHYLENE-INDUCED METABOLIC CHANGES ASSOCIATED WITH RESISTANCE REACTIONS
Phytoalexins and enzymes

The possible involvement of ethylene in plant defenses against invading pathogens has attracted considerable attention (5). Ethylene can induce both enzymes implicated in resistance reactions and phytoalexins, anti-microbial secondary metabolites which form part of a greater activation of new plant metabolism, which occurs as a consequence of stress, particularly following microbial infections and after treatment with toxic compounds or toxic microbial metabolites (3). Thus, ethylene induces the antifungal compound 3-methyl-6-methoxy-8-hydroxy-3,4-dihydro-

isocoumarin (MMHD) in carrot disks in a concentration- and exposure time-
dependent manner (7, 12). Similarly, ethylene has been demonstrated to
induce the phytoalexins pisatin in pea stem tissue (11), and phytuberin
and phytuberol in tobacco (40). The enzyme phenylalanine ammonia-lyase
(PAL) has been implicated in the biosynthesis of phytoalexins in pea
tissue (19) and found to be stimulated by ethylene in tissues from sweet
potato, citrus, cucumber (1, 25), carrot (10) and tobacco (33). However,
in the case of cut tissue from white potato, ethylene failed to influence
PAL induction (21), although it stimulated the induction of phytoalexins
after elicitation by several fungi (20).

Stimulation of PAL activity is not necessarily dependent upon increased
ethylene production. In soybean infected with Phytophthora megasperma var.
sojae ethylene can be uncoupled from stimulation of PAL and phytoalexin
production. Like the fungus itself, an eliciting cell wall preparation
induced the synthesis of ethylene 1.5 h, PAL 3 h and the phytoalexin
glyceollin 6 h after application. AVG inhibited the induced ethylene
formation and ACC substituted for the elicitor as far as ethylene production was concerned. AVG treatment did not inhibit, nor did ACC treatment
stimulate either PAL of glyceollin formation (29). Thus, whereas the
productions of PAL and glyceollin were linked, those of ethylene and PAL
were not, indicating that the stimulation of ethylene, on the one hand,
and of PAL and phytoalexin, on the other hand, proceed by different
routes.

Ethylene has been described to stimulate many hydrolytic and oxidative
enzymes (1, 47). In particular polyphenoloxidase and peroxidase (37) have
been implicated in the production of quinones toxic to the pathogen and
the formation of barrier substances (e.g. lignins) confining the pathogen
to the site of penetration. However, it is not clear as to how far stimulation of these enzymes can act as a determinant of resistance. In wheat
infected with stem rust, Puccinia graminis, leaves from a resistant line
upon ethylene treatment became completely susceptible, in spite of the
fact that high peroxidase levels were induced (13). In tobacco, virus infection stimulates polyphenoloxidase and peroxidase activity independent
of symptom type, be it at different times after inoculation and in hypersensitively reacting plants enzyme activity seems to be a reflection of
a physiological state rather than being responsible for regulating the
rate of virus spread (42, 46).

Usually, ethylene production increases most when symptoms are already apparent and the resulting changes in enzyme activities may occur too late to be involved in the initial defense reaction. However, they may still play a role in the gradual build-up of a mechanism preventing extensive spread of the pathogen from its original site of penetration. An early, high rise of ethylene production at the infection site will stimulate such responses and has been linked to the expression of a resistance reaction against various pathogens: the induction of chitinase in several plant species, the enhancement of the synthesis of cell wall hydroxyproline-rich glycoprotein in melon infected with Colletotrichum lagenarium, and reduction of virus spread in TMV-infected tobacco.

Chitinase

Chitin does not occur in plants but is a component of fungal and bacterial cell walls. The induction of chitinase by ethylene (2) could thus be a defense reaction initiated when plant tissues are damaged, and directed against fungal or bacterial invaders. In bean leaves, chitinase activity increased 30-fold within 24 h of treatment with 10 ppm ethylene, and in fully induced leaves constituted more than 1% of the total protein (6). Continuous presence of ethylene was necessary for full induction. Induction was inhibited by both cycloheximide and AVG. The purified enzyme readily attacked cell walls of the potential pathogenic fungus F. solani and acted as a lysozyme on bacterial cell walls. Experiments with melon seedlings inoculated with C. lagenarium (38), with pea pods infected with F. solani and with tobacco leaves reacting hypersensitively to TMV showed that chitinase was also induced during pathogenesis. An enhanced synthesis of ethylene accompanied the plant's response in all these cases. However, treatment of the infected pea pods with AVG did not block the induction of chitinase, although it prevented stress ethylene formation. Conversely, ACC stimulated ethylene formation strongly but caused only a small increase in chitinase activity. The enzyme $\beta-1,3-$glucanase showed similar characteristics in this tissue. Thus, at least in pea pods, these enzymes were induced by the fungus independently of ethylene (28).

Hydroxyproline-rich cell wall glycoprotein

In C. lagenarium-infected melon stems and petioles, an inverse relationship has been found between the accumulation of hydroxyproline-rich glycoprotein of the cell walls and the ability of the pathogen to develop in the host. Ethylene increased the amount of the hydroxyproline-rich glycoprotein in the wall, while at the same time increasing resistance of the host to the pathogen (17). In infected plants, AVG reduced both ethylene production and the incorporation of ^{14}C-hydroxy-proline into the cell walls, whereas treatment of healthy plants with ACC stimulated both ethylene production and ^{14}C-hydroxyproline incorporation (39). Although the effects of both AVG and exogenously applied ethylene were rather small in comparison to those induced by the fungus, they are indicative of at least a stimulative effect of ethylene on this defensive response.

Pathogenesis-related proteins

During the hypersensitive reaction of tobacco to TMV several new proteins accumulate up to 10% of the total leaf protein. These pathogenesis-related proteins (PRs) are induced after infection with viruses, fungi or bacteria, notably when hypersensitive or systemic necrosis occurs. Non-necrotizing viruses may induce PRs in relatively small amounts at later stages of infection, particularly when bright yellow symptoms are prominent. Thus, necrosis or wounding is not a prerequisite for PR appearance. The presence of PRs is linked to reduced multiplication and spread of the pathogen; even under conditions where the pathogen is confined to the inoculated leaf, PRs are induced throughout the entire plant associated with the development of systemic acquired resistance (41, 45).

Both the induction of PRs and the development of systemic acquired resistance can be induced in noninfected plants by treatment of a few leaves on the plant with ethephon (44). Only moderate induction occurs when ethephon is applied as a spray or rubbed onto the leaves. However, by mimicking viral lesion formation by pricking leaves with needles moistened with high concentrations of ethephon, inductions similar to those achieved by virus infection may be obtained. The stimulation of peroxidase activity and changes in the isozymic pattern provoked by the virus are likewise simulated by ethephon pricking (Fig. 1). PRs can also be induced by incubating leaf discs on ACC and when TMV-inoculated leaf discs are incubated on AVG, the amount of PRs produced is strongly reduced. In general, there is a good correlation between the increase in ethylene

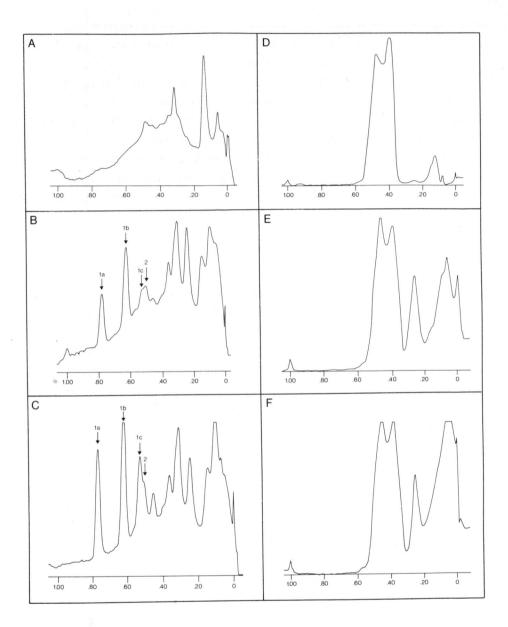

FIGURE 1. Densitometer tracings of electrophoretic patterns in 10 and 7.5% polyacrylamide gels, respectively, of pH 3-soluble proteins (A-C) and anionic peroxidases (D-F) from Nicotiana tabacum cv. Samsun NN, untreated (A,D) , or 7 days after pricking with needles moistened with 0.3 M ethephon (B,E) or inoculation with 10 mg TMV per 1 (C,F). Pathogenesis-related proteins in (A-C) are designated as 1a, b, c and 2. After challenge inoculation of upper leaves from duplicate plants with TMV, relative lesion sizes were 100, 53, and 51, respectively.

production and the amount of PRs induced. Since the appearance of local
lesions in hypersensitively reacting tobacco is accompanied by a large
burst of ethylene after which PRs start to accumulate, ethylene appears
to function as the physiological inducer after virus infection and probably other pathological conditions where PRs become apparent.

Apart from their association with reduced virus multiplication and
spread, thus far no specific function of PRs has been established. The
occurrence of PRs and similar proteins in pathological conditions is a
wide-spread phenomenon (32). PRs from bean share several distinguishing
properties with those of tobacco: they are a family of proteins of relatively low molecular weight, differing mainly in isoelectric point; they
are extractable at low pH, highly resistant to endogenous and foreign
proteases and localized predominantly in the intercellular space. Recent
evidence indicates that each of the four most prominent PRs in tobacco
is translated from a separate mRNA (8). These mRNAs are not newly trans-
cribed upon stimulation, because when mRNA prepared from healthy leaves
was translated in a cell-free protein-synthesizing system, as much PRs
were synthesized as from mRNA from infected leaves. These observations
demonstrate that PR-mRNA is constitutively present but not translated in
non-stimulated leaves (9). The effect of ethylene might thus consist of
making the mRNAs available for translation. Since ethylene is produced
by most tissues as a general reaction to stress (47), PRs might thus have
a function in protecting the plant from extensive damage.

REFERENCES

1. Abeles FB. 1973. Ethylene in plant biology. Academic Press, New York.
2. Abeles FB, Bosshart RP, Forrence LE, Habig WH. 1970. Plant Physiol.
   47, 129-134.
3. Bailey JA. 1983. In: The dynamics of host defence (Bailey JA, Deverall
   BJ, eds), pp. 1-32. Academic Press, Sydney.
4. Bailiss KW, Balázs E, Király Z. 1977. Acta Phytopathol. Acad. Sci.
   Hung. 12, 133-140.
5. Boller T. 1982. In: Plant Growth Substances 1982 (Wareing PF, ed.),
   pp. 303-312. Academic Press, London.
6. Boller T, Gehri A, Mauch F, Vögeli U. 1983. Planta 157, 22-31.
7. Carlton BC, Peterson CE, Tolbert NE. 1961. Plant Physiol. 36, 550-552.
8. Carr JP. 1983. Neth. J. Plant Pathol. 89, in press.
9. Carr JP, Antoniw JF, White RF, Wilson TMA. 1982. Biochem. Soc. Trans.
   10, 353-354.
10. Chalutz E. 1973. Plant Physiol. 51, 1033-1036.
11. Chalutz E, Stahmann MA. 1969. Phytopathology 59, 1972-1973.
12. Chalutz E, De Vay JE, Maxie EC. 1969. Plant Physiol. 44, 235-241.
13. Daly JM, Seevers PM, Ludden P. 1970. Phytopathology 60, 1648-1652.

14. De Laat AMM, Van Loon LC. 1982. Plant Physiol. 69, 240–245.
15. De Laat AMM, Van Loon LC. 1983. Physiol. Plant Pathol. 22, 261–273.
16. De Laat AMM, Van Loon LC, Vonk CR. 1981. Plant Physiol. 68, 256–261.
17. Esquerré-Tugayé MT, Lafitte C, Mazau D, Toppan A, Touzé A. 1979. Plant Physiol. 64, 320–326.
18. Gáborjányi R, Balázs E, Király Z. 1971. Acta Phytopathol. Acad. Sci. Hung. 6, 51–56.
19. Hadwiger LA, Hess SL, Von Broemssen S. 1970. Phytopathology 60, 332–336.
20. Henfling JWDM, Lisker N, Kuč J. 1978. Phytopathology 68, 857–862.
21. Hyodo H, Yang SF. 1974. Z. Pflanzenphysiol. 71, 76–79.
22. Kamerbeek GA, De Munk WJ, 1976. Sci. Horticult. 4, 101–115.
23. Kato S. 1976. Ann. Phytopathol. Soc. Jpn 43, 587–589.
24. Levy D, Marco S. 1976. Physiol. Plant Pathol. 9, 121–126.
25. Lisker N, Cohen L, Chalutz E, Fuchs Y. 1983. Physiol. Plant Pathol. 22, 331–338.
26. Marco S, Levy D. 1979. Physiol. Plant Pathol. 14, 235–244.
27. Marco S, Levy D, Aharoni N. 1976. Physiol. Plant Pathol. 8, 1–7.
28. Mauch F, Hadwiger LA, Boller T. 1984. Plant Physiol., in press.
29. Paradies I, Elstner EF. 1980. Ber. Deutsch. Bot. Ges. 93, 635–657.
30. Pritchard DW, Ross AF. 1975. Virology 64, 295–307.
31. Purohit AN, Tregunna EB, Ragetli HWJ. 1975. Virology 65, 558–564.
32. Redolfi P. 1983. Neth. J. Plant Pathol. 89, in press.
33. Reuveni M, Cohen Y. 1978. Physiol. Plant Pathol. 12, 179–189.
34. Ross AF. 1961. Virology 14, 340–358.
35. Ross AF, Pritchard DW. 1972. Phytopathology 62, 786.
36. Sitterly WR, Shay JR. 1960. Phytopathology 50, 91–93.
37. Stahmann MA, Clare BG, Woodbury W. 1966. Plant Physiol. 41, 1505–1512.
38. Toppan A, Roby D. 1982. Agronomie 2, 829–834.
39. Toppan A, Roby D, Esquerré-Tugayé MT. 1982. Plant Physiol. 70, 82–86.
40. Uegaki R, Fujimori T, Kaneko H, Kubo S, Kato K. 1980. Phytochemistry 19, 1543–1544.
41. Van Loon LC, 1975. Virology 67, 566–575.
42. Van Loon LC. 1976. Physiol. Plant Pathol. 8, 231–242.
43. Van Loon LC. 1976. Abstr. 9th Int. Conf. Plant Growth Substances, Lausanne, pp. 412–414.
44. Van Loon LC. 1977. Virology 80, 417–420.
45. Van Loon LC. 1983. In: The dynamics of host defence (Bailey JA, Deverall BJ, eds), pp. 123–190. Academic Press, Sydney.
46. Van Loon LC, Geelen JLMC. 1971. Acta Phytopathol. Acad. Sci. Hung. 6, 9–20.
47. Yang SF, Pratt HK. 1978. In: Biochemistry of wounded plant tissues (Kahl G, ed.), pp. 595–622. De Gruyter, Berlin.

ETHYLENE BIOSYNTHESIS IN TOBACCO LEAF DISCS IN RELATION TO ETHYLENE TREAT-
MENT, CELLULYSIN APPLICATION AND FUNGAL INFECTION

Edo Chalutz[1], Autar K. Mattoo[2] and James D. Anderson[2]

Agricultural Research Organization, The Volcani Center, Bet Dagan, Israel[1]
and Plant Hormone Laboratory, BARC, ARS, USDA, Beltsville, Maryland, USA[2]

INTRODUCTION

A common feature of many plant diseases is an increase in ethylene
production (1,18). In addition, many pathogenic microorganisms produce
ethylene when grown in culture (1,13,14,18). Based on these observations,
a role for ethylene in disease development has been postulated (5,6,18,22).
However, little is known about the mechanism of ethylene production during
infection, or about the site of ethylene production, or the relative contribution of host or pathogen to the ethylene produced. Possibly ethylene production is associated with early events following the action of enzymes
secreted by the pathogen during disease development since involvement of
cell-wall degrading enzymes is known to constitute an early event in the
interaction of host and pathogen (4,23). Indeed, a cell-wall digesting
preparation of fungal origin, "Cellulysin", was reported (3,8,9) to induce
ethylene biosynthesis in tobacco leaf discs by causing a rapid formation
of ACC, the immediate precursor of ethylene in higher plants (25). This
process is stimulated by prior exposure of the tissue to ethylene (8,9).
We have characterized this phenomenon, studied its relationship to fungal
infection and the role of ethylene in regulating this process.

MATERIALS AND METHODS

Three cultivars of tobacco (Nicotiana tabacum L.) plants were used
viz. Burly Mammoth, Maryland 609 and Xanthi. Materials and methods used
for the preparation of leaf discs and ethylene determination were described
earlier (3).

Tobacco leaves were pretreated in air or ethylene in 3.8-L desiccators.
Each leaf was divided in half by cutting along its mid-rib and each half
was placed on filter paper, moistened with water, in individual desiccators.
A vial containing filter paper soaked with 2 ml of 0.25 M mercuric perch-

Y. Fuchs and E. Chalutz (eds.) Ethylene: Biochemical, Physiological and Applied Aspects.
ISBN 90-247-2984-X. Printed in The Netherlands

lorate was placed in the "air control" desiccator to absorb traces of ethylene. Unless otherwise indicated, 3 leaf discs (1 cm in diameter, weighing
50 mg) were incubated in 25-ml Erlenmeyer flasks with 0.5 ml of the basal
medium (3). Cellulysin (Calbiochem) was desalted before use by ultrafiltration with an Amicon PM-10 membrane (2). Ethylene was allowed to accumulate
for 1 h and quantified by gas chromatography (15). Between each sampling,
flasks were flushed with sterile fresh air. Other methods and materials
have been described in detail in an earlier publication (8).

     Three replicates were routinely used and experiments were repeated
at least twice and gave similar pattern of results. However, due to the
variability of the greenhouse-grown plants, results of typical experiments
are presented.

RESULTS

     Tobacco leaf discs incubated in a medium containing Cellulysin respond
by increased ethylene production. This response of leaf discs to Cellulysin
was further enhanced several-fold by pretreating the detached leaf in ethylene (Table 1). However, when leaf discs instead of whole leaves were pretreated in air, their subsequent response to Cellulysin was similar to that
of the discs cut from ethylene-treated leaves. Ethylene-treated leaf discs
produced ethylene in response to Cellulysin at a rate 1.5 times higher than
discs treated in air (Table 1). These rates of ethylene production are
among the highest rates reported for any biological system.

Table 1. The effect of different pretreatments on the Cellulysin-induced
        ethylene production by tobacco (cv. Burly Mammoth) leaf discs.

Treatment	Ethylene production
	nl/g/h
Freshly cut (no pretreatment)	160
Pretreated as leaf (halves)	
air	185
C_2H_4	550
Pretreated as discs	
air	495
C_2H_4	760

Discs without Cellulysin addition produced ethylene at rates lower than
45 nl/g fresh wt/h. Reference 8.

Of the tobacco cultivars studied, Burly Mammoth produced the highest
rates of ethylene in response to Cellulysin; cultivar differences occurred
whether freshly cut leaves or leaves pretreated in ethylene were used,
although some cultivars (i.e. TI 102) did not respond to the ethylene pretreatment.

The maximal response of the leaf to ethylene pretreatment was reached
between 4 and 8 h of incubation. Periods longer than 10 h of incubation
did not further increase the subsequent response of the discs to Cellulysin.
The optimal concentration of ethylene for maximal response was between
10 and 100 μl/l.

To verify that the increased production of ethylene by the discs from
ethylene-pretreated leaves in the presence of Cellulysin was due to de
novo synthesis instead of release of absorbed or bound ethylene from the
tissue, discs from leaves pretreated in ethylene were incubated with (3,4-
^{14}C) methionine and Cellulysin. Total and labeled ethylene were then assayed. The results (Fig. 1) show that Cellulysin-induced ethylene production
in ethylene-treated tissue resulted from the conversion of methionine to
ethylene. Furthermore, these data showed close similarities in the pattern
of total and labeled ethylene produced and in the specific radioactivity
of ethylene produced by Cellulysin-treated discs from freshly cut or ethylene-pretreated leaves, suggesting a common biosynthetic pathway.

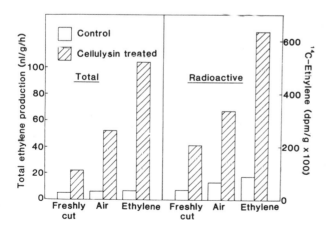

Figure 1. Comparison bet-
ween rates of total and
labeled ethylene produc-
tion from (3,4-^{14}C) methionine by tobacco (cv.
Xanthi) lead discs. The
values represent average
rates of ethylene produc-
tion during the first 2
h of incubation after the
addition of Cellulysin.
Reference 8.

Cellulysin caused a marked accumulation of ACC in the treated tissue
(Table 2), a phenomenon observed earlier (3). However, this effect was
pronounced in the freshly cut or air-pretreated discs and was relatively
inconspicuous in discs pretreated in ethylene (Table 2).

Table 2. Total and labeled ACC content and ethylene produced by tobacco
(cv. Xanthi) discs incubated with $(3,4-^{14}C)$ methionine in relation
to different pretreatments.

Leaf tissue treatment	Cellulysin addition	C_2H_4 $g^{-1}h^{-1}$	ACC	
		(nmoles)	(dpm x 10^{-3})	(nmoles)
Fresh	-	0.15	3.80	1.40
	+	0.65	14.30	26.95
Pretreated in air; 16h	-	0.20	3.50	1.05
	+	1.50	7.30	30.10
Pretreated in C_2H_4; 16h	-	0.20	3.45	1.05
	+	3.25	2.85	7.70

Leaf discs, either freshly cut or after the indicated treatments given
to whole leaves, were incubated with the basal medium (6 discs/ml) containing 1952 dpm/nmol methionine. Adapted from ref. 8.

The lower content of ACC in the ethylene-treated discs in the presence
of Cellulysin could result from a higher rate of conversion of ACC to ethylene. To test this possibility, we compared the rates of ethylene production
by freshly cut leaf discs as well as by discs cut from air or ethylene
pretreated tissue in response to exogenously added ACC (without Cellulysin)
while blocking the formation of endogenous ACC by including 1 mM AVG in the
incubation medium. The results (Table 3) confirmed previous observations
in Nicotiana (16) and other systems (12, 19) that aging of leaf tissue
increases their capacity to produce ethylene from ACC. Our data further
indicated that the conversion of exogenously added ACC to ethylene was
twice as much in ethylene-treated than in air-treated leaf tissue while
the uptake of labeled ACC by the ethylene-treated discs was lower than
that by the air-treated or freshly cut leaf discs.

Table 3. The effect of pretreatment of tobacco (cv. Xanthi) leaves in ethylene on their subsequent conversion of ACC to ethylene, in the absence of Cellulysin. Values of ethylene production are for the 1st h of incubation following the addition of ACC (1.0 mM), in the presence of 1mM AVG.

Treatment	Ethylene production	
	Control	ACC-treated
	(nl/g/h)	
Freshly cut	10	33
Preincubated in air	11	42
Preincubated in ethylene	11	74

Tobacco leaf discs inoculated with the pathogen, A. alternaria (21), produced ethylene at higher rates than did uninoculated controls (8,9). Moreover, inoculated discs cut from leaves pretreated in ethylene produced higher rates of ethylene than those cut from fresh tissue or from leaves pretreated in air. Also, A. alternaria produced ethylene when cultured on a low phosphate containing medium (data not shown).

DISCUSSION

The mode of action of ethylene in enhancing the response of tobacco leaves to Cellulysin may be complex. The following findings suggest an explanation: (a) pretreatment of leaf discs in air could partially substitute the effect of pretreatment in ethylene of whole leaves (Table 1) and (b) pretreatment of leaf tissue in air also slightly, but consistently, enhanced the response of the tissue to Cellulysin (Table 1,Fig. 1) as compared to freshly cut leaves. On this basis we suggest that ethylene could be involved in the enhancement of the tissues' response at very low concentrations (i.e. less than 8 nl/l). Alternatively, these findings may suggest that other factors, in addition to ethylene, may be involved in the process. Such a suggestion was offered by Geballe and Galston who studied wound-induced resistance to cellulase in oat leaves (11) and reported that ethylene was a factor in this process (10). A similar phenomenon was observed earlier in prune tissue (24). However, since exogenously applied ethylene could only partially substitute for the wounding effect, Geballe and Galston suggested (10) that the induction of resistance may require a wound signal in addition to ethylene.

The results presented here clearly indicate that higher rates of ethylene production are not merely the result of the release of ethylene from the tissue. Rather, it originates from enhanced biosynthesis from methionine (Fig. 1). Also, these data indicate that in all leaf treatments, i.e. freshly cut, air and ethylene pretreated, a common precursor and biosynthetic pathway lead to the production of ethylene (Fig. 1). However, following incubation with Cellulysin, the ACC content was much lower in discs from ethylene-treated leaves than from freshly cut or air pretreated discs (Table 2). We interpret these findings to suggest that while Cellulysin induced the formation of ACC, presumably due to increased activity of ACC synthase (3), ethylene treatment of leaves causes a higher rate of conversion of ACC to ethylene, thus increasing the utilization of newly formed ACC in the ethylene-treated tissue. This suggestion is supported by the findings that discs from ethylene-treated leaves converted exogenously applied ACC to ethylene at a higher rate that discs from freshly cut or air-pretreated leaves (Table 3).

In some respects, the effect of ethylene reported in citrus leaf discs by Riov and Yang (20), and a similar observation reported recently for preclimacteric canataloupe (12) resembles the enhancement by ethylene of the Cellulysin-induced ethylene production reported here. However, unlike the requirement of 36 to 48 h by citrus leaf discs to exhibit increased ACC formation and ethylene production in response to ethylene, the enhanced response of tobacco leaf to ethylene as exhibited by measuring Cellulysin-induced ethylene production, occurs within 4 to 8 h.

Also, in our studies the faster response of the tissue to ethylene was on the ethylene-forming enzyme and not on ACC synthesis. This delineation was possible since ACC synthesis is induced by Cellulysin within 1 h of incubation (3) and thus is not rate limiting.

Since the ethylene forming enzyme seems to be membrane-associated (17,25) and ethylene is known to cause changes in membrane permeability (1), it is possible that increased conversion of ACC to ethylene in ethylene-treated tissue may be mediated through a change in the membrane milieu of this enzyme.

In considering the host-pathogen relationships, it is attractive to formulate a model wherein a fungal attack promotes the formation and accumulation of ACC, possibly by some kind of a fungal elicitor. Concomitantly, host tissue responds by enhancing the endogenous formation of ethylene from ACC in an autocatalytic mode leading to a rapid production of ethylene. Consequently, senescence of leaf tissue is enhanced, thus, predisposing it to the advancing fungal mycelium. Alternatively, the induction of increased production of ethylene could be part of a process by which the plant maintains a defense mechanism aimed at combating infections. This could be brought about by induction of phytoalexins (7,18) or by making the cell-wall less susceptible to pathological degradation (10,11).

ACKNOWLEDGEMENT

Supported in part by the U.S.-Israel Binational Agricultural Research & Development Fund (BARD).

LITERATURE CITED

1. Abeles FB. 1973. Ethylene in plant biology. Academic Press, London, New York.

2. Anderson JD, Lieberman M, Stewart RN. 1979. Ethylene production by apple protoplasts. Plant Physiol 63: 931-935.

3. Anderson JD, Mattoo AK, Lieberman M. 1982. Induction of ethylene biosynthesis in tobacco leaf discs by cell wall digesting enzymes. Biochem Biophys Res Commun 107: 588-598.

4. Bateman DF, Basham HG. 1976. Degradation of plant cell walls and membranes by microbial enzymes. In Physiological Plant Pathology, (R Heitefuss and PH Williams, eds) pp 316-355, Springer Verlag Berlin, Heidelberg and New York.

5. Boller T. 1982. Ethylene-induced biochemical defenses against pathogens. In Plant Growth Substances (PF Wareing ed) pp 303-312, Academic Press, London.

6. Boller T, Mauch GF, Vogeli V. 1983. Chitinase in bean leaves: Induction by ethylene, purification, properties and possible function. Planta 156: 22-31.

7. Chalutz E, DeVay JE, Maxie EC. 1969. Ethylene-induced isocoumarin formation in carrot root tissue. Plant Physiol 44: 235-241.

8. Chalutz E, Mattoo AK, Solomos T, Anderson JD. 1984. Enhancement by ethylene of Cellulysin-induced ethylene production by tobacco leaf discs. Plant Physiol 74: 99-103.

9. Chalutz E, Mattoo AK, Anderson JD. 1983. Cellulysin-induced ethylene production by tobacco leaf discs in relation to ethylene produced during host-pathogen interaction. Proc Plant Growth Reg Soc America pp 18-24.

10. Geballe GT, Galston AW. 1982. Ethylene as an effector of wound-induced resistance to cellulase in oat leaves. Plant Physiol 70: 788-790.

11. Geballe GT, Galston AW. 1982. Wound-induced resistance to cellulase in oat leaves. Plant Physiol 70: 781-787.

12. Hoffman NE, Yang SF. 1982. Enhancement of wound-induced ethylene synthesis by ethylene treatment in preclimacteric cantaloupe. Plant Physiol 69: 317-322.

13. Ilag L, Curtis RW. 1968. Production of ethylene by fungi. Science 159: 1357-1358.

14. Lieberman M. 1979. Biosynthesis and action of ethylene. Ann Rev Plant Physiol 30: 533-591.

15. Lieberman M, Kunishi AT, Mapson LW, Wardale DA. 1966. Stimulation of ethylene production in apple tissue slices by methionine. Plant Physiol 41: 376-382.

16. Mattoo AK, Lieberman M. 1982. Role of silver ions in controlling senesence and conversion of 1-aminocyclopropane -1-carboxylic acid to ethylene. Plant Physiol Suppl 69: 18.

17. Mattoo AK, Achilea O, Fuchs Y, Chalutz E. 1982. Membrane association and some characteristics of the ethylene forming enzyme from etiolated pea seedlings. Biochem Biophys Res Commun 105: 271-278.

18. Pegg CF. 1976. The involvement of ethylene in plant pathogenesis. In Physiological Plant Pathology (R Heitefuss and PH Williams eds) pp 582-591 Springer Verlag Berlin, Heidelberg and New York.

19. Rhodes MJC. 1980. The maturation and ripening of fruits. In KV Thimann ed. Senescence in Plants. CRC Press Boca Raton, FL pp 157-205.

20. Riov J. Yang SF. 1982. Effect of exogenous ethylene on ethylene production in citrus leaf tissue. Plant Physiol 70: 136-141.

21. Spurr HW Jr. 1973. An efficient method for producing and studying tobacco brown-spot disease in the laboratory. Tobacco Sci 17: 145-148.

22. Stahmann MA, Clare BG, Woodbury W. 1966. Increased disease resistance and enzyme activity induced by ethylene and ethylene production by black rot infected sweet potato tissue. Plant Physiol 41: 1505-1512.

23. Suzuki K, Furusawa I, Ishida N, Yamomoto M. 1982. Chemical dissolution of cellulose membranes as a prerequisite for penetration from appresoria of Collectotrichum lagenarium. J Gen Microbiol 128: 1035-1039.

24. Weinbaum SA, Labavitch JM, Weinbaum Z. 1979. The influence of ethylene treatment of immature prune (Prunus domestica L.) fruit on the enzyme-mediated isolation of mesocarp cells and protoplasts. J Am Soc Hortic Sci 104: 278-280.

25. Yang SF. 1980. Regulation of ethylene biosynthesis. Hort Sci 15: 238-243

PURIFICATION AND PROPERTIES OF THE ETHYLENE-INDUCING FACTOR FROM THE CELL WALL DIGESTING MIXTURE, CELLULYSIN[1]

James D. Anderson[2], Edo Chalutz[3] and Autar K. Mattoo[4], Plant Hormone Laboratory,[2,4], BARC, ARS, Beltsville, Maryland, Volcani Research Center[3], Bet Dagan, Israel, and Botany Department[4], University of Maryland, College Park, Maryland, USA.

INTRODUCTION

The production of ethylene during development and senescence of higher plants is a well established fact. Also, stress (including physical and chemical wounding), or application of herbicides and hormones, as well as host-pathogen interaction induce ethylene production (11). Recently we reported a new type of ethylene-inducing agent, a cell wall digesting enzyme mixture, 'Cellulysin' (4). The induction of ethylene in tobacco leaves (4) by Cellulysin can be relatively fast, within 30 minutes, compared to that by the hormone IAA which takes hours (2). Cellulysin-mediated induction of ethylene biosynthesis is inferred to be at the level of ACC synthase because ACC is produced and 0.1mM aminoethoxyvinylglycine (AVG), a known inhibitor of this enzyme (1), inhibits both the induction of ACC and ethylene biosynthesis. The mechanism by which Cellulysin induces ethylene biosynthesis is not known, but it occurs even in the presence of the protein synthesis inhibitors, cycloheximide (50µM) and chloramphenicol (100 µg/ml).

In our attempt to further understand the induction of ethylene biosynthesis by Cellulysin, the need arose to purify the component(s) responsible for it. Cellulysin is known to be a mixture of many

[1]This work was carried out in part under the Cooperative Agreement No. 58-32U4-1-216 of the United States Department of Agriculture, Agricultural Research Service and the University of Maryland and supported in part by a grant from the United States-Israel Binational Agricultural Research and Development Fund (BARD).

We thank Tommie Johnson, Marcia Sloger, and Annette Thomas for assistance. Mention of a company name or trademark does not constitute endorsement by the U.S. Department of Agriculture over other of a similar nature not mentioned.

Y. Fuchs and E. Chalutz (eds.) Ethylene: Biochemical, Physiological and Applied Aspects.
ISBN 90-247-2984-X. Printed in The Netherlands
©1984, Martinus Nijhoff/Dr W. Junk Publishers, The Hague.

enzymes (eg. 6,13). Here we report isolation, purification and some properties of the component(s) from Cellulysin that is responsible for inducing ethylene biosynthesis in plant tissues.

PROCEDURES

Plant material.

Tobacco plants (Nicotiana tabacum L. cv. Xanthi and Burly Mammoth) were grown in a greenhouse with natural lighting. During the summer months temperature in the greenhouse exceeded 40C on some days.

Bioassay of ethylene biosynthesis induction. In most experiments, leaf discs were used. Ethylene biosynthesis did not reach its maximum rate as fast in Burly Mammoth as it does in Xanthi (8), and therefore the times of analysis were not always the same. In all experiments, leaves were treated overnight in 60-100 µl/l ethylene in order to increase the amount of ethylene produced by leaf discs in response to Cellulysin (8,9 and as reported by Chalutz, et al in a preceding paper).

Separation techniques. Cellulysin (from Calbiochem) was desalted by ultrafiltration with a Amicon PM-10 membrane (3). The desalted enzyme was used in all subsequent experiments. Desalted Cellulysin was fractionated on 6ml columns of Sephacryl-200 S and Sephadex G-100 using 10 mM MES, pH 6.0, as the equilibration buffer and on a discontinuous sucrose isoelectric focusing column, pH 3-10 (5). Other enzymes were desalted on 6ml Sephadex G-25 columns.

Stability characteristics. In some experiments a partially purified fraction was characterized as regards its stability to temperature and proteolytic enzymes. Aliquots of the partially purified fraction (after Sephacryl-200 and isoelectric focusing) were incubated for 10 minutes at various temperatures prior to bioassaying the residual activity or treated with 90 µg/ml of each protease for 1 hr at 30C prior to bioassay. Cellulysin (2.45 mg protein) was incubated with 500 µg SDS in a volume of 100 µl for 2 h at 30C prior to bioassaying 100 µg of Cellulysin protein.

Electrophoresis. Sodium dodecyl sulfate polyacrylamide gel electrophoresis (SDS-PAGE) on 10 and 12% acrylamide gels followed by staining with a modified silver stain (12) or double staining (10) with Coomassie brilliant blue and the modified silver stain were used to

determine protein patterns in several enzyme preparations used in
protoplast isolation.

RESULTS AND DISCUSSION

The induction of ethylene biosynthesis is not unique to the enzyme
mixture, Cellulysin. Several other enzyme mixtures used for protoplast
release and isolation are also capable of inducing ethylene, some as
effective as Cellulysin (Table I). Cellulase containing preparations
closely related to Cellulysin were quite active as was pectinase,
Rhozyme, Driselase and Pectolyase. Other enzymes (e.g., Cellulases P, B
and PB from Worthington, macerase and a hemicellulase) were ineffective
in inducing ethylene biosynthesis at least at the concentration tested,
100 µg protein/ml. Our initial attempts to associate the active
component with some protein in the different enzyme preparations, to the
ethylene inducing component using SDS-PAGE gel fractionation were not
successful. As can be seen in Figure 1 all enzyme mixtures were quite
complex and no specific protein could be related to the ability of the
enzyme mixture to induce ethylene production. A more methodical approach
was needed.

Initially, Sephracryl-200 S columns were used to fractionate the
activity, but problems arose with recovery and in some experiments
activity seemed to bleed off the column long after any added component
should have remained. This is demonstrated in Table 2 using mini

Table 1. Induction of ethylene in tobacco (cv. Burly Mammoth) leaf
discs by cell-wall digesting enzyme preparations during the 5th h of
incubation with 100µg protein/ml of desalted enzyme.

Enzyme	Source	Ethylene Production Freshly cut	Ethylene-treated
		nl/g.hr	
Control		13	14
Cellulysin	Calbiochem, USA	120	625
Cellulase (RS)	Yakult Pharm. Ind. Co., LTD	42	860
Cellulase	Sigma, USA	16	66
Pectinase	Sigma, USA	23	117
Rhozyme	Rohm and Hass, USA	16	53
Driselase	Kyowa Hakko Kogyo Co., LTD	13	44
Pectolyase	Seishin Pharm. Co., LTD	61	131

Other enzymes tested that gave little or no ethylene-inducing response
at 100µg/ml protein were cellulases P,B, and PB (Worthington); Macerase
(Calbiochem) and Hemicellulase (Sigma).

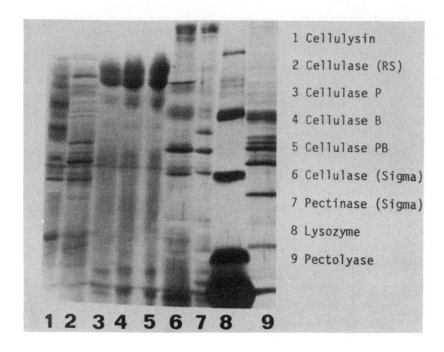

1 Cellulysin

2 Cellulase (RS)

3 Cellulase P

4 Cellulase B

5 Cellulase PB

6 Cellulase (Sigma)

7 Pectinase (Sigma)

8 Lysozyme

9 Pectolyase

1 2 3 4 5 6 7 8 9

Figure 1. A Commassie blue-silver stained 12% SDS polyacrylamide gel of different enzyme preparations used in protoplast isolation.

columns (volume 6ml) of Sephacryl-200 S. Sephacryl gave a good purification and the specific activity increased greatly (i.e., up to 50-fold). It is evident from this data that some activity is lost and probably remains on the column. The material that interacts with Sephacryl and comes off late does not just represent small molecules because the material did not pass through an Amicon PM 10 membrane, which should have retained molecules greater than 10,000 M.W. Including 50mM NaCl in the column buffer, or increasing the pH to 8, did not prevent this apparent interaction or binding of the active component to Sephacryl. The interaction between the active component and Sephacryl does not show up in Sephadex-G100 where the activity seems to remain with the protein fraction.

Table 2. Apparent interaction between the ethylene inducing component
from Cellulysin and Sephacryl-200 S

| Fraction | Sephadex G-100 | | | Sephacryl-200 S | | |
	Activity	Protein	Sp. Act	Activity	Protein	Sp. Act
	units[1]	mg	units/mg	units	mg	units/mg
High M.W.	13.9	2.6	5.3	n.d	2.3	-
Intermediate M.W.	10.9	5.6	1.9	2.7	4.0	0.7
Low M.W.	10.9	2.0	5.5	9.7	1.0	9.7
Interacting A	n.d.[2]	0.2	-	2.9	-	120.0
Interacting B	n.d.	n.d.	-	1.6	-	66.7
Total recovery	35.7	10.4		16.9	7.3	
Recovery (%)	152	104		72	73	

Cellulysin (10 mg protein) applied to each column would induce 23.5
μl/g.hr in tobacco (cv. Xanthi) leaf discs during the 2nd - 3rd hr of
incubation (Sp.Act. = 2.4 units/mg protein).
[1]Units = μl of ethylene induced per g tissue during the 2nd to 3rd hr
of incubation.
[2]Not detected.

The ethylene inducing activity of Cellulysin did not focus in a
narrow pH range when subjected to discontinuous sucrose gradient
isoelectric focusing (Fig. 2). Instead, activity spread over a wide pH
range. The focusing of the activity over a wide pH range could
indicate, among other possibilities, a multitude of charged molecules or
a large molecule (e.g. protein) with various modifications.

By using a combination of the above methods we have obtained a very
active component, presumably a protein, but the concentration of the
purified component was low and below the limit of detectability of the
Bradford assay (7).

The partially purified component is inactivated 60% in 10 minutes at
60C and 100% at 70C. The ethylene-inducing activity of Cellulysin is
lost when the preparation is incubated with low levels (0.5%) of SDS
(Table 3). Thus, it has some properties that are associated with
proteins, but various proteases do not inhibit the action of this
component. Attempts are being made to accumulate a large quantity of
the purified, active component, in order to better characterize it.

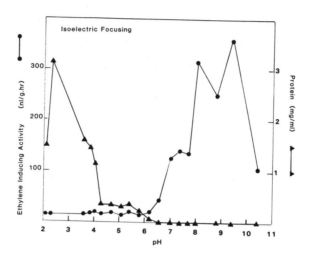

Figure 2. Distribution of ethylene inducing activity of Cellulysin on a isoelectric focusing column.

One of the major problems we faced initially was the fact that some batches of our bioassay plants did not respond to Cellulysin even after ethylene pretreatment (Table 4). The tobacco plants that did not respond, including both Xanthi and Burly Mammoth cultivars, were started from seed during the summer when temperatures in the greenhouse exceeded 40C during the heat of the day. Plants started earlier in the spring and that experienced the same greenhouse conditions maintained their responsiveness to Cellulysin. It does not appear that greenhouse conditions per se were the cause for nonresponsiveness but it might be more related to the growing conditions when seeds were germinated or transplanted. Interestingly, the non-responsive plants were affected almost to the same degree as the responsive plants by ethylene in that

195

Table 3. Sensitivity of the Cellulysin, ethylene-inducing factor to
SDS, proteases and temperature.

Treatment	Response
SDS	At 0.5%, SDS inhibited ethylene inducing capacity of Cellulysin >50% at 30C for 2 hr.
Proteases	Ethylene inducing activity of a partially purified fraction[1] was not affected significantly by a 1 h incubation at 30C in the presence of 90µg/ml proteinase K, trypsin, chymotrypsin, protease V8 from _Staphylococcus aureus_ or a combination of the four proteases.
Temperature	Ethylene inducing activity of a partially purified fraction[1] was inhibited 0%, 60% and 100% when incubated for 10 minutes at 50C, 60C or 70C, respectively, prior to bioassay.

[1] Active fraction was partially purified by a combination of
membrane ultrafiltration, Sephacryl-200 chromatography, and isoelectric
focusing.

the _in vivo_ conversion of ACC to ethylene was stimulated (Table 4) as
previously reported (9). Thus, the cause of the unresponsiveness of
these plants is not because the plants do not respond to ethylene, but
that they do not respond to the active component of Cellulysin in
stimulating ACC synthase activity. The reason for the unresponsiveness
is strictly speculative at this time. The unresponsive plants seemed to
remain unresponsive to Cellulysin even after the plants were cut back
and allowed to grow out again (unpublished).

Interestingly, the responsive plants pretreated with ethylene
produce more ethylene in response to Cellulysin than they do when
supplemented with exogenous ACC. The greater increase in ethylene
production in ethylene-pretreated leaves was shown previously (9) not to
be related to ACC uptake differences. The mechanism by which ethylene
pretreatment increases the conversion of ACC to ethylene is under
investigation.

To determine if tobacco is unique in its response to Cellulysin we
surveyed a number of different species to determine how widespread this
phenomena is and to see if other species would be better than tobacco
for bioassay. It is clear from the results presented in Table 5 that

Table 4. Differential responsiveness of Burly Mammoth tobacco plants to ethylene induction with Cellulysin.

Treatment	Less Responsive Plants	Responsive Plants
	nl/g.hr	nl/g.hr
Freshly cut control	21	12
" " + Cellulysin	18	26
" " + 3×10^{-5}M ACC	77	59
" " + 1×10^{-4}M ACC	79	103
" " + 3×10^{-4}M ACC	188	135
Ethylene pretreated control	29	15
" " + Cellulysin	44	900
" " + 3×10^{-5}M ACC	200	250
" " + 1×10^{-4}M ACC	364	414
" " + 3×10^{-4}M ACC	550	456

leaf tissue from other species do indeed respond to Cellulysin by producing ethylene as do some fruit tissues. However, some species did not respond in these experiments. At this time, we do not intend to infer that corn and sugarbeet leaves do respond to the active component of Cellulysin because the plants used in these studies might simply reflect a situation similar to what we already discussed about some tobacco plants that don't respond to the active component of Cellulysin. These experiments do show that tobacco is not unique and that other species do respond to Cellulysin by producing ethylene. From the species tested, tobacco seemed to be the better bioassay plant because of the size of leaf and the ease by which it is grown.

SUMMARY

Various cell wall digesting enzyme mixtures used for protoplast isolation were shown to induce ethylene production in tobacco leaf discs. This induction does not appear to be related to destruction of the cell wall or other obvious wounding phenomena. In fact, purified cellulases did not induce ethylene. A minor component of Cellulysin, that appears to be a protein was partially purified by a combination of

Table 5. Cellulysin-induced ethylene production by several plant tissues. All tissues were pretreated with ethylene as described for tobacco.

Plant Material	Relative ethylene production
Leaf tissue	
Tobacco (Burly Mammoth)	+ + +
Cotton	+ + +
Rape	+ +
Spinach	+ +
Lettuce	+ +
Cabbage	+ +
Corn	--
Sugar beet	--
Fruit tissue	
Apple	+ +
Orange (peel)	+ +
Banana	+ (?)
Tomato	+ (?)
Avocado	- (?)
Potato	- (?)

membrane ultrafiltration, Sephacryl chromatography and isoelectric focusing. Attempts are being pursued to further purify the component in order to further characterize it, and to raise antibodies against it. The antibodies should help to locate where the active component interacts with cells in order to induce ethylene. Such studies should lead to our understanding of how a macromolecule, believed to be proteinaceous, can interact with plant cells to induce ethylene.

REFERENCES

1. Adams DO, Yang SF. 1979. Ethylene biosynthesis: Identification of 1-aminocyclopropane-1-carboxylic acid as an intermediate in the conversion of methionine to ethylene. Proc. Natl. Acad. Sci., USA 76:170-174.
2. Aharoni N, Anderson JD, Lieberman M. 1979. Production and action of ethylene in senescing leaf discs. Effect of indoleacetic acid, kinetin, silver ion, and carbon dioxide. Plant Physiol. 64: 805-809.
3. Anderson JD, Lieberman M, Stewart RN. 1979. Ethylene production by apple protoplasts. Plant Physiol. 63:931-935.
4. Anderson JD, Mattoo AK, Lieberman M. 1982. Induction of ethylene biosynthesis in tobacco leaf discs by cell wall digesting enzymes. Biochem. Biophys. Research Commun. 107: 588-596.
5. Baker, JE. 1976. Superoxide dismutase in ripening fruits. Plant Physiol. 58: 644-647.

6. Boller T, Kende H. 1979. Hydrolytic enzymes in the central vacuole of plant cells. Plant Physiol. 63: 1123-1132.
7. Bradford M. 1976. A rapid and sensitive method for the quantitation of microgram quantities of protein utilizing the principle of protein-dye binding. Anal. Biochem. 72: 248-254.
8. Chalutz E, Mattoo AK, Anderson JD. 1983. Cellulysin-induced ethylene production by tobacco leaf discs in relation to ethylene produced during host-pathogen interactions. Proceedings Plant Growth Regulator Soc. Amer. 18-24.
9. Chalutz E, Mattoo AK, Solomos T, Anderson JD. 1984. Enhancement by ethylene of Cellulysin induced ethylene production by tobacco leaf discs. Plant Physiol 74: 99-103.
10. Eschenbruch M, Burk RR. 1982. Experimentally improved reliability of ultrasensitive silver staining of protein in polyacrylamide gels. Anal. Biochem 125:96-99.
11. Lieberman M. 1979. Biosynthesis and action of ethylene. Ann. Rev. Plant Physiol. 30: 533-591.
12. Moline HE, Johnson KS, Anderson JD. 1983. Evaluation of two-dimensional polyacrylamide gel electrophoresis of acidic proteins of ribosome preparations for identifying plant pathogenic soft-rotting bacteria. Phytopathology 73: 224-227.
13. Saunders JA. 1979. Investigations of vacuoles isolated from tobacco. I. Quantitation of nicotine. Plant Physiol 64: 74-78.

THE INVOLVEMENT OF CALLOSE AND ELICITORS IN ETHYLENE PRODUCTION CAUSED
BY MECHANICAL PERTURBATION.

M. J. JAFFE.

Biology Dept., Wake Forest University, Winston-Salem, N.C. USA

When plants are mechanically perturbed (MP) by wind, rubbing, flexing or other means, they undergo thigmomorphogenesis (10). This reaction,
which is ubiquitous, usually consists of a retardation of elongation of
the stem coupled with a thickening of the stem (10). Both terrestrial and
aquatic plants are subject to perturbation by fluid flow: The former by
wind and the latter by currents of water (25). It is probable that
thigmomorphogenesis arose in response to fluid flow in the environment.
In earlier papers (10,13,14), I have proposed the thigmomorphogenetic
hypothesis: namely that mild or moderate MP causes plants to be more
resistant to rupture due to subsequent more severe MP. Thus, if a stand
of trees are sheltered from the wind by the perimeter trees, and the
perimeter trees are cut down, the sheltered trees will be susceptible to
stem rupture (falling) by the first strong wind that occurs (9). This
hypothesis now has enough experimental support to be called the
"Thigmomorphogenetic Theory" (17,22). The present paper will describe
some of these data, together with what is known of the hormonal mechanism
of thigmomorphogenesis.

THE HORMONAL BASIS OF THIGMOMORPHOGENESIS

Several species will be discussed in this paper, however, the
primary experimental species will be the garden bean (Phaseolus vulgaris,

Y. Fuchs and E. Chalutz (eds.) Ethylene: Biochemical, Physiological and Applied Aspects.
ISBN 90-247-2984-X. Printed in The Netherlands

L., cv. Cherokee Wax) and loblolly pine (Pinus taeda, L.). The
thigmomorphogenesis induces both stunting and thickening in these two
species. Different kinds of MP produce much the same results, and Tab. 1
shows that rubbing or flexing of the stem provides a good model system
for the kind of wind induced thigmomorphogenesis that the plants would
experience in nature. Accordingly, MP was given to the bean plants by
rubbing the stems (11) and to the pine plants by flexing them (22,23).

TABLE 1. The effects of different amounts of MP due to wind, flexing
or rubbing on thigmomorphogenesis.

Treatment	Length of First Internode (mm)	Diameter of First Internode (mm)
None (Control)	60 ± 1	2.8 ± 0.1
Wind (10, 5m/sec Gusts of 10 sec ea/day)	36 ± 2	3.5 ± 0.1
Flexing (5 x @ 30°)	36 ± 2	3.7 ± 0.1
Rubbing (5 x per day)	21 ± 1	3.4 ± 0.1

If one examines the gnarled and swollen appearance of MP-treated bean
or pine stems (13,22), it is apparent that they resemble organs that
have been treated with ethylene. Indeed, if non-perturbed plants are
exogenously treated with ethephon, they appear similar to those which
have undergone thigmomorphogenesis (Tab. 2). This similarity can also
be seen on the anatomical level. In beans both exogenous ethephon and
MP induce lateral cell divisions in the secondary xylem (2,13), decreases
in cell elongation in the cortical cells (2) and thickening of the
epidermal cutical (13). In pine, either MP or ethephon causes decreases
in tracheid length and increases in the number of tracheids (22,23).

If ethylene is causally involved in thigmomorphogenesis, ACC and
ethephon should mimic the effects of MP and inhibitors of ACC (AVG) and
ethylene ($CoCl_2$) biosynthesis should block the morphological reactions

to MP. That such is exactly the case is shown in Tab. 3.

TABLE 2. Morphological changes due to MP or exogenous ethephon in bean and loblolly pine.

Plant		Control	MP	mm Ethephon
Bean	First Internode Length (mm)	80	69	59
	Diameter (mm)	2.4	2.9	3.8
Pine	Hypocotyl Length (mm)	110	87	94
	Diameter (mm)	3.4	3.7	3.8

TABLE 3. The effects of phytohormones and ethylene inhibitors on thigmomorphogenesis.

Additive		Bean L	dia	Pine L	dia
None (C)[*]		−50	+40	−21	+ 9
IAA (10uM)[**]		−36	−37	−	−
GA (1uM)[**]		+27	− 7	−	−
ABA (1uM)[**]		−24	−30	−	−
ACC [**]		−44	+65	−	−
Ethephon [**]		−44	+52	−15	+12
AVG [***]		−51	+ 4	−	−
CoCl$_2$ [***]		−46	− 3	−	−

[*] Effect of MP (%Δ)

[**] Effect of the additive on growth on non-MP plants (%Δ)

[***] Effect of the additive on thigmomorphogenesis (%Δ).

It whould be noted that whereas AVG and CoCl$_2$ completely inhibit the thickening response to MP they have no effect on the stunting response. This suggests that ethylene directly mediates the thickening component of thigmomorphogenesis, but only indirectly mediates the decrease in elongation. In order to explain the stunting response, the action of other

202

phytohormones must be invoked (5,21). Tab. 3 shows that high concentrations of IAA, and a low concentration of ABA both cause a decrease in elongation of the bean stem. Both of these phytohormones cause the stem to become thinner than that of controls. Furthermore, a low concentration of GA overcomes the stunting effect of MP.

If these other phytohormones are causally involved in thigmomorphogenesis, their content should change after MP. Tab. 4 shows that the native titre of both IAA and ABA increases following MP, whereas that of GA decreases. Thus, the stunting effect of MP may be due to the disappearance of GA, a hormone responsible for elongation, an increase in ABA, a hormone known to be a natural growth retardant, and an increase in IAA to levels above 1uM, which retard growth.

Ethylene, however, is the direct mediator of the thickening response of thigmomorphogenesis. Tab. 4 shows that in both bean and pine, MP causes an increase in ethylene evolution. Genetic experiments are demonstrated in the table, as well. The pine half-sibling 8-27 which evolves MP-induced

TABLE 4. The effect of MP on the hormone content of bean and pine stems.

Plant	Hormone	No. of MP days before sampling	Hormone Content (% Δ)
Bean	Ethylene	1	+54
	IAA	10	+70
	ABA	10	+68
	GA[1]	29	-57
Pine			
1/2 sib 8-27	Ethylene	90	+123
1/2 sib 8-64	Ethylene	90	-90
Tomato			
cv Hosen	Ethylene	1	+11
cv Alcobaca	Ethylene	1	+153

[1] After suge 1980.

ethylene, is thickened by MP while the half-sibling 8-61, by which less

ethylene is evolved following MP, is not thickened. Similarly, the Hosen

tomato cultivar evolves little MP-induced ethylene and only thickens 3%

due to MP. However, MP induces 50% thickening in the Alcobaca cultivar,

which also evolves a great deal of MP-induced ethylene. The ethylene that

is produced as a result of MP represents an increase in "endogenous"

ethylene but not in "wound" ethylene (Huberman and Jaffe, unpublished).

In both pine and bean, MP-induced ethylene evolution begins after lag

periods of about 3 and 0.5 h, respectively, peaks at about 9 and 2 h,

respectively, and decays away after 50 and 5 h, respectively (Figures 1

and 2). Thus, ethylene evolution, due to MP, in both species occurs as

a pulse following a substantial lag time. It is this pulse of ethylene

production that appears to be directly responsible for the thickening

reaction of thigmomorphogenesis.

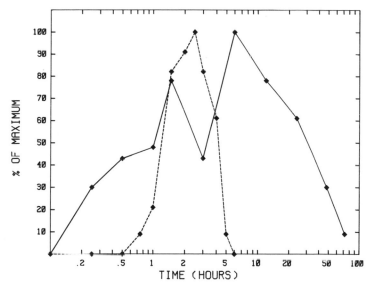

FIGURE 1. The time courses of MP induced (MP-C) callose deposition (——)
and ethylene production (- - -) by bean first internodes.

THE ROLE OF CALLOSE

But, if there is a period of time between which MP is perceived by
the plant, and ethylene begins to evolve, then there is an apparent
causal chain linking the two. My laboratory has been interested in the
nature of the possible links of that chain. It is possible that one of
the links is the cell wall polysaccharide, callose. Callose is a probably
cross linked high polymer composed of glucose units linked together with
B (1,3) bonds, instead of the B (1,4) bonds found in cellulose. It
apparently is deposited outside of the plasma membrane whenever a plant
receives a trauma. Rapid callose deposition has been anecdotally reported
to occur following handling of plants (7), and we have recently characterized the MP-induced deposition in bean stems (15). The function of
callose is unknown, but it may act, in some plants, and under certain
circumstances to play a role in blocking basipetal phloem transport of
photosynthate, such as we have been able to show in beans (Fig. 3). In
this experiment, the basipetal transport of radioactively labeled photosynthate is rapidly inhibited by MP below the rubbed region, and accumulates above the rubbed region (Fig. 3).

Whatever its physiological functions may turn out to be, callose
deposition certainly is correlated to thigmomorphogenesis. Figures 1 & 2
show that callose deposition begins well before ethylene evolution in
both bean and pine. In fact, in the case of bean, its deposition commences
immediately after MP. Thus, on the basis of timing, callose deposition
may be involved in the mediation of MP-induced ethylene evolution. In
addition, another correlation may be found by probing with 2-deoxy-D-
glucose (DDG), an inhibitor of protein glycosylation. This inhibitor
blocks callose deposition in bean and pine, and also stops thigmomorphogenesis in bean and ethylene production in bean and pine (Tab. 5).

TABLE 5. The effects of an inhibitor of protein glycosylation (DDG) on thigmomorphogenesis and callose and ethylene production. The data represent the effects of the DDG on MP-plants (%Δ).

Measurement	Bean	Pine
Length	+137	-
Diameter	- 84	-
Callose deposition	- 77	-24
Ethylene evolution	- 54	-34

TABLE 6. The effect of various carbohydrates on callose deposition or ethylene production in bean stem tissue.

Treatment	% of Control C_2H_4 production (nl/g/h)	Callose Deposition (μm)
None control	100 ± 0	100 ± 5
MP	176 ± 2	-
UDPG (10mM)	135 ± 5	352 ± 11
Laminarin (0.5%)	173 ± 6	-9 ± 8
Starch (0.5%)	107 ± 5	-
Carboxy methyl cellulose (0.1%)	102 ± 3	-
Glucose (0.5%)	105 ± 3	-

However, if this involvement is causal, callose would have to be shown to induce ethylene evolution, itself. Table 6 shows the results of several experiments designed to test this hypothesis. Of all the additives (made up in a 10 mM MES buffer, pH 5.8), only UDP-glucose (the precursor of callose biosynthesis), and laminarin (a commercially available callose derived from the alga Laminaria) were able to induce ethylene evolution when challenging bean stem tissue. Neither glucose itself, cellulose (B(1,4) glucan), nor starch (A(1,6) glucan) enduced ethylene evolution.

Since callose is deposited on the outside of the plant cell proto-

plast and on the inside of the plant cell wall, it is in the right
location to affect any enzymes at the cell surface. The enzyme system
that converts ACC to ethylene has been suggested to occur at the cell
surface (18). Thus, it may be that MP induces callose deposition, and
that the peculiar conformation of the B(1,3) glucan linkages stimulate
ethylene biosynthesis from ACC at the cell surface.

Laminarin represents about 5% of the total cell wall callose in
Laminaria, but it is soluble in water and therefore useful for the kind
of experiments in hand. However, the methods available for the extraction and purification of callose from higher plants, produce an insoluble
form of callose. In order to prove that native callose is involved in
the induction of ethylene evolution, it will be necessary to challenge
bean stems (for example) with callose extracted from bean stems. We are
currently attempting to devise a procedure for extracting pure, water
soluble callose from this tissue.

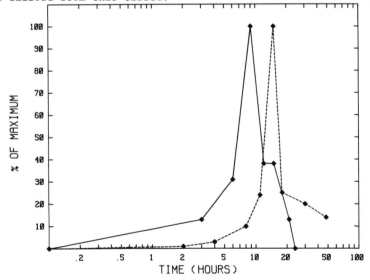

FIGURE 2. The time courses of MP induced (MP-C) callose deposition (———)
and ethylene production (— — —) by loblolly pine hypocotyls.

THE ROLE OF ELICITORS

In the process of trying to understand the mechanisms of thigmomorphogenesis we found that when one part of the plant is mechanically perturbed, other parts react as well. Thus, for example, if the first internode is rubbed, the second and third internodes also undergo thigmomorphogenesis (4,11). This phenomenon also extends to biochemical changes (5,
6). It seems, therefore, that some messenger can carry MP information
across graft unions (4). This messenger does not seem to be ethylene,
since ethylene is not known to move within plants, or IAA, since it moves
acropetally. It cannot be GA, since stunting is the result, or a cytokinin
since they have not been shown to mimic MP. Two possible candidates for
the messenger are ABA and ACC. Neither of these possibilities have been
ruled out; however, we have also investigated another possibility involving elicitors.

Elicitors are molecules, produced by a plant or its pathogen, which
elicit the production of stress metabolites. Among the stress metabolites
are phytoalexins, which are toxic to the pathogen and confer resistance
to it in the potential host. However, elicitors are also produced by
plants in response to abiotic stresses and stress metabolites are produced
by treatment with exogenous ethylene (8). Elicitor activity has been shown
to be found in 12 unit oligosaccharides of galacturonic acid (19), in B
(1,3) oligosaccharides (16), and in fatty acids such as arachidonic acid
(3). Since all of these types of molecules might be expected to be found
in the extracellular space, we have modified the method of Terry and his
co-workers (24) to elute putative elicitors from bean stem tissue (Fig.4).
The elicitor-like activity, measured by the soybean cotyledon bioassay of
Albersheim and Valent (1), is greater in eluates of MP stems than from
controls (Tab. 7). Furthermore, it is capable of inducing somewhat more

ethylene evolution than is the control eluate (Tabl. 7). Perhaps, of the most interest, however, is the fact that when MP eluate is exogenously applied to bean stems, it induces the same morphological changes as does MP (Tab. 7). Inhibitors of ethylene biosynthesis, such as AVG or $CoCl_2$ have a stron inhibitory effect on the eluted elicitor-like activity, but early elicitor activity does not induce ethylene evolution in bean stem tissue. The time course of MP-induced elicitor-like activity is interesting. Figure 5 shows that there is an initial rapid increase in elicitor-like activity, followed by a rapid decrease by 3 h. The MP-induced elicitor-like activity increases by a moderate amount at 5 to 6 h, and remains at or near that level for at least 24 h. This later MP-induced elicitor-like activity is completely destroyed when the eluate is treated with pectinase. Therefore, it seems possible that the elicitor may be the oligosaccharide composed of units of galacturonic acid described by Nothnagel et al. (19). The possible relationship of MP-induced elicitor to ethylene is obscure, at present. However, it seems clear that the elicitor is capable, in and of itself, of mediating thigmomorphogenesis in beans.

TABLE 7. The effect of control - and MP-eluates on stem thickening, ethylene production and elicitor-like activity (glyceolin production, measured at 285 nm in the soybean cotyledon bioassay).

Measurement	Distilled H_2O	Control Eluate	MP Eluate	MP
Stem thickening (% of initial growth)	2 ± 1	4 ± 1	9 ± 1	9 ± 1
Ethylene evolution (ul/g/h)	0.56 ± 0.05	0.72 ± 0.01	1.02 ± 0.17	1.22 ± 0.01
Elicitor activity (A285 nm)	0.283	0.426	0.475	–

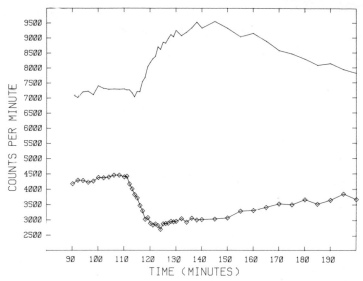

Figure 3. The effect of brief rubbing a previously unperturbed stem on basipetal transport of $^{11}CO_2$ photosynthate above the MP region (——) and in the subtending hypocotyl (◆—◆). The MP was given between 111 & 112 min. The previous level of radioactivity represents the pre-MP baseline.

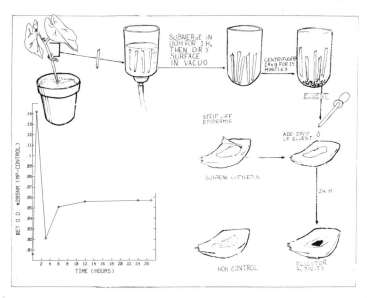

Figure 4. The extraction and assay of elicitor-like substances from MP bean stems. The inset graph shows the time course of MP-induced (MP-C) elicitor-like activity. The MP plants were rubbed at 0 and 24 h. Extracts were made at intervals and applied to the soybean cotyledon bioassay which was read after 24 hours.

In summary, the hormonal mechanism of thigmomorphogenesis in bean and perhaps in pine, can be said to include direct mediation of the thickening reaction by ethylene-induced lateral cell divisions, and separately perhaps by an elicitor. The stunting reaction may also be mediated by the elicitor, and probably also involves a decrease in native gibberellins and an increase in native IAA and ABA. With this understanding of the mechanisms of thigmomorphogenesis, we may now address the question of whether or not MP induces resistance to mechanical rupture in plants.

CONFIRMATION OF THE THIGMOMORPHOGENETIC THEORY

According to the thigmomorphogenetic theory, slight or moderate winds (or other MP) will harden the plant to withstand mechanical rupture by subsequent strong winds. Table 8 shows that both pine and bean plants undergo thigmomorphogenesis due to rubbing or flexing. If, after thigmomorphogenesis has occurred, a lateral force load (simulating wind) is applied to the plants, they break less easily than do controls. As shown in the table, bean is 65%, and pine is 55% more resistant to mechanical rupture than controls. Thus, in both of these plants, the thigmomorphogenetic theory is upheld. The biomechanical reason for this resistance, however is different in the two species. In both species the modulus (E) is about the same, indicating an increase in stem flexibility. However, in the case of the pine plants, there is an increase in the flexural stiffness (EI). In other words, thigmomorphogenesis induces hardening in the pine stems by making them stiffer, so that they will not bend and then break. On the other hand, it causes hardening in the bean stems by making them so flexible that they will bend a great deal without breaking. It is interesting, but not at all surprising that different mechanisms of MP-induced hardening have evolved in different plant species. It is no less significant that the same type of hardening can be induced by exogenous treatment with ethephon, and that it is due to the same type of changes in the biomechanical properties of the stems.

TABLE 8. The effects of MP on various mechanical properties of bean stems and loblolly pine hypocotyls.

Parameter	$\frac{C-MP}{C}$ x 100	
	Bean	Pine
Length (mm)	-32	-6
Diameter (mm)	+ 8	+34
Plants broken under a force load (% of total)	-65	-55
Elastic modulus (E)	-39	-36
2D Moment of Area (I)	+53	+109
Flexural Stiffness (EI)	-30	+18

It is significant that, in addition to hardening a plant against mechanical rupture, MP also may harden it against other types of stress induced injury. Examples of this phenomenon are shown in Table 9. For example, both pine and bean that have undergone thigmomorphogenesis are more resistant to frost stress than are control plants (13). Similarly, previously mechanically perturbed bean or tomato plants are hardened against drought stress-induced growth inhibition or pithiness, respectively (13,20). In the latter case, at least, we have been able to show that exogenous ethephon can substitute for MP in making the tomato stems less susceptible to drought induced pithiness (20).

Pithiness is the autolysis of pith tissue and seems to be very widespread among plants. In bean, stem pithiness seems to occur as part of the natural development of the stem (Tab. 10). However, it is greatly retarded by MP, which also

TABLE 9. The effects of MP on resistance to other stresses by bean or loblolly pine stems.

Plant	Treatment	MP Advantage $(\frac{MP-C}{C} \times 100)$
Bean	Withold irrigation, then measure new growth 5 days after re-irrigation (wt. of new growth in g)	114%
Tomato	Withold irrigation, then measure pithiness following re-irrigation (% Pithiness)	-94%
Bean	Expose to $-20^{\circ}C$ for 5 min, then incubate 4 days at $26^{\circ}C$ and measure new growth (mm)	145%
Pine	Expose to overnight frost, then allow to grow for 1 season and compute % of plants with live candles (%)	18%

TABLE 10. The effect of phytohormones, ethylene inhibitors and environmental stresses on ethylene production and pithiness of bean first internodes.

Treatment	Ethylene Production (ul/g/h)	Pithiness Holc (% of cross sectional area)
None (control)	1.5 ± 0.2	8 ± 1
Drought stress	3.4 ± 0.6	1 ± 1
MP	2.0 ± 0.2	3 ± 1
Ethrel	-	1 ± 0
IAA (2×10^{-6}M)	2.8 ± 0.2	4 ± 1
GA (2×10^{-5}M)	-	8 ± 1
ABA (2×10^{-7}M)	-	5 ± 1
Kinetin (2×10^{-5}M)	2.6 ± 0.2	4 ± 1
AOA (2×10^{-4}M)	0.6 ± 0.1	9 ± 0
Kinetin + AOA	0.9 ± 0.1	6 ± 1

induces an increase in ethylene production. We have characterized the responses of this system to various phytohormones, and can show that ethylene, kinetin or IAA can impart protection against the development of pithiness. Furthermore, both IAA and kinetin seem to produce their results by inducing an increase in ethylene production (Table 10).

Thus, it seems that the effect of wind and other types of mechanical perturbations, can be expressed in a number of ways. On the morphological level, stems are stunted and thickened, and their mechanical properties are considerably altered. This alteration can be expressed ecologically in that it renders the plants less susceptible to mechanical rupture (by e.g., high winds), and thus better able to successfully pass through its life cycle. In addition, it also may make the plant more resistant to certain other environmental stresses. All of these considerations strongly suggests that thigmomorphogenesis is beneficial to the plant, rather than harmful, and represents a syndrome which provides an adaptive advantage to the plant.

ACKNOWLEDGEMENTS:

I would like to thank the following students, Post-Doctoral fellows and Visiting Scientists who have contributed to various aspects of this work: Dr. Y. Erner, Dr. F. Telewski, Dr. M. Huberman, Dr. R. Biro, Dr. H. Takahashi and Dr. E. Pressman. Various aspects of this work were supported by research grants from the National Science Foundation (USA), the National Aeronautical and Space Administration (USA) and the Binational Research and Development Program (USA and Israel).

214

REFERENCES

1. Albersheim, P. and Valent, B. S. 1978. Host-pathogen interaction in plants: Plants, when exposed to oligosaccharides of fungal origin, defend themselves by accumulating antibiotics J. Cell Biol. 78: 627-643.
2. Biro, R., Hunt, E. R., Jr. and Jaffe, M. J. 1980. Thigmomorphogenesis: Changes in cell division and elongation in the internodes of mechanically perturbed or ethrel treated bean plants. Ann. Bot. 45: 655-664.
3. Bostock, R. M., Kuc, J. A. and Laine, R. A. 1981. Eicosapentaenoic and arachidonic acids from phytophtera infestens elecit fungitoxic sesquiterpenes in the potato Science. 212: 67-69.
4. Erner, Y., Biro, R. and Jaffe, M. J. 1980. Thigmomorphogenesis: Evidence for a translocatable thigmomorphogenetic factor induced by mechanical perturbation of beans (Phaseolus vulgaris). Physiol. Plant. 50: 21-25.
5. Erner, Y. and Jaffe, M. J. 1982. Thigmomorphogenesis. The involvement of auxin and abscisic acid in growth retardation due to mechanical perturbation. Plant and Cell Physiol. 23: 935-941.
6. Erner, Y. and Jaffe, M. J. 1983. Thigmomorphogenesis: Membrane lipid changes in bean plants as affected by mechanical perturbation and Ethrel Physiol. Plant. 58: 197-203.
7. Eschrich, W. 1975. Sealing systems in plants. In Vol. I (I) of Encyclopedia of Plant Physiology pp. 39-56. M. H. Zimmerman and J. A. Milburn, Eds., Springer-Verlag, New York.
8. Haard, N. F. 1983. Stress metabolites in post harvest physiology and crop preservation Ed. M. Lieberman. pp. 299-314. Plenum Publishing Co.
9. Jacobs, M. R. 1954. The effect of wind sway on the form and development of Pinus radiata D. Don. Australian J. Bot. 2: 35-51.
10. Jaffe, M. J. 1973. Thigmomorphogenesis. The response of plant growth and development to mechanical stimulation: with special reference to Bryonia Dioica. Planta 114: 143-157.
11. Jaffe, M. J. 1976. Thigmomorphogenesis: A detailed characterization of the response of beans (Phaseolus vulgaris L.) to mechanical stimulation Z. Pflanzenphysiol. 77: 437-453.
12. Jaffe M. J. and Biro, R. 1977. Thigmomorphogenesis: The role of ethylene in wind induced growth retardation. Proc. Plant Gr Reg Work Gp 4: 118-124.
13. Jaffe, M. J. 1979. Thigmomorphogenesis. Anatomical changes, the role of ethylene and interaction with other environmental stresses. In: Stress physiology in Crop Plants. Ed. H. Mussel and R. C. Staples. pp. 25-59, Wiley, New York.
14. Jaffe, M. J. 1980. Morphogenetic responses of plants to mechanical stimuli or stress. Bioscience 30(4): 239-243.
15. Jaffe, M. J., Huberman, M. Johnson, J. and Telewski, F. W. Thigmomorphogenesis: The induction of callose and ethylene by mechanical perturbation in bean stress. Physiol. Plant. In press.
16. Keen, N. T. and Yoshikawa, M. 1983 β-1,3-endoglucanase from soybean releases elicitor - active carbohydrates from fungus cell walls. Plant Physiol. 71: 460-465.
17. Lawton, R. O. 1982. Wind stress and elfin stature in a montane rain forest tree: An adaptive explanation. Amer. J. Bot. 69: 1224-1230.

18. Mattoo, A. K. Achilea, D., Fuchs, Y. and Chalutz, E. 1982. Membrane association and some characteristics of the ethylene forming enzyme from etiolated pea (Pisum sativum cv. Calvedon) seedlings Biochem. Biophys. Res. Comm. 105: 271-278.

19. Nothnagel, E. A., McNeil, M. Albersheim, P. and Dell, A. 1983. Host-pathogen interactions XXII. A galacturonic acid oligosaccharide from plant cell wall elicits phytoalexins. Plant Physiol. 71:916-926.

20. Pressman, E., Humberman, M., Aloni, B. and Jaffe, M. J. 1983. Thigmomorphogenesis: The effect of mechanical perturbation and ethrel on stem pithiness in tomato (Lycopesicon esculentum, Mill.) plants Annals of Botany 52: 93-100.

21. Suge, H., 1978. Growth and gibberellin production in Phaseolus vulgaris as affected by mechanical stress. Plant and Cell Physiol. 19: 1559-1560.

22. Telewski, F. W., 1983. On the mechanism of thigmomorphogenesis in conifers. Doctoral Dissertation. Wake Forest University. Winston-Salem, N.C., U.S.A.

23. Telewski, F. W. and Jaffe, M. J. 1981. Thigmomorphogenesis. Changes in the morphology and chemical composition induced by mechanical perturbation in 6 month old Pinus taeda seedlings. Can. J. For. Res. 11(2): 380-387.

24. Terry, M. E. and Bonner, B. A. 1980. An examination of centrifugation as a method of extracting an extracellular solution from peas, and its use for the study of indoleacetic acid induced growth. Plant Physiol. 66: 321-325.

25. Vogel, S. 1981. Life in moving fluids. Willard Grant Press, Boston. ISBN 0-87150-749-8. 352 pp.

ELICITORS AND ETHYLENE TRIGGER DEFENSE RESPONSES IN PLANTS

M.T. ESQUERRÉ-TUGAYÉ, D. MAZAU, B. PELISSIER, D. ROBY, A. TOPPAN

Université Paul Sabatier, Centre de Physiologie Végétable, 118, Route
de Narbonne, 31062 Toulouse, France

INTRODUCTION

Infected plants often produce large amounts of ethylene. The possible
role of this hormone in plant-microorganism interaction is poorly understood.
We have previously demonstrated that ethylene protects melon plants against
Colletotrichum lagenarium , the causal agent of anthracnose (1). This has
led to the hypothesis that ethylene could be a message which triggers
defense responses in plants. The increased synthesis of proteins associated
to defense mechanisms during plant-microorganism interactions, and their
elicitation by ethylene, or via ethylene by fungal elicitors, are
successively reported. The 3 proteins considered in this study are cell
wall hydroxyproline-rich glycoprotein (HRGP, which strengthens the plant
cell wall), chitinase (which digests chitin-containing fungal cell wall),
and proteolytic inhibitors (which neutralize the trypsin-like proteases
of plant pathogens).

MATERIAL AND METHODS

Plant material : Melon seedlings (~ 21 day-old) inoculated (I) with
C. lagenarium, and healthy (He) controls (1) ; excised He leaves or petioles.

Elicitors : Ethanol-soluble fragments recovered from a crude preparation
obtained by autoclaving fungal mycelium or plant cell walls. Excised plant
material received elicitors (2) either by immersion (petioles), or
absorption (leaves).

Ethylene : was determined by G.C.

Proteins correlated to defense responses : HRGP was measured either
colorimetrically, or by scintillation counting of ^{14}C-Hydroxyproline
released by HCl hydrolysis of the cell walls of plants incubated with
^{14}C-Pro (3). Chitinase activity was colorimetrically assessed (4).
Proteolytic inhibitors were measured through their ability to inhibit the
proteolysis of azocoll by trypsin or C. lagenarium protease.

Y. Fuchs and E. Chalutz (eds.) Ethylene: Biochemical, Physiological and Applied Aspects.
ISBN 90-247-2984-X. Printed in The Netherlands
©1984, Martinus Nijhoff/Dr W. Junk Publishers, The Hague.

RESULTS

The melon-Colletotrichum lagenarium system has been used throughout these studies.

Ethylene production. Melon seedlings infected by C. lagenarium start producing increased amounts of ethylene 30 hours after inoculation (4). Elicitors either of fungal (C. lagenarium) or of plant (melon) origin have the ability to mimic the plant pathogen interaction, i.e. to induce the synthesis of ethylene in healthy melon tissues. Elicitation of ethylene takes place 60 to 90 minutes after the beginning of the treatment. The ethylene response of protoplasts to elicitors is delayed by several hours.

Effect of ethylene, of elicitors of ethylene, and of infection, on the level of HRGP, chitinase, and proteolytic inhibitors. The levels of the three proteins are highly increased upon infection of melon plants by Colletotrichum lagenarium. They are also increased after treatment of healthy seedlings or tissues with ethylene or with ACC, and after treatment with elicitors. When these treatments are performed prior to inoculation, melon susceptible plants tend to become systemically protected against C. lagenarium.

Is ethylene a message in the plant defense ? Ethylene is increased in infected plants 1 to 2 days before the 3 proteins under study ; inhibition of its synthesis by AVG results in the simultaneous inhibition of HRGP synthesis. AVG also inhibits the elicitor promoted accumulation of HRGP, chitinase, and proteolytic inhibitors.

CONCLUSION

Ethylene, and elicitors of ethylene, trigger defense responses in plants. Their partial inhibition by AVG supports the hypothesis that ethylene is one of the intermediary messages which induce the plant defense.

REFERENCES

1. Esquerré-Tugayé MT, Lafitte C, Mazau D, Toppan A, Touzé A. 1979. Plant Physiol. 64, 320-326.
2. Roby D, Toppan A, Esquerré-Tugayé MT. 1984. Plant Physiol. (submitted).
3. Toppan A, Roby D, Esquerré-Tugayé MT. 1982. Plant Physiol. 70, 82-86.
4. Toppan A, Roby D. 1982. Agronomie 2, 829-834.

THE ROLE OF ETHYLENE IN THE PATHOGENIC SYMPTOMS DISPLAYED BY <u>MELOIDOGYNE JAVANICA</u> NEMATODE INFECTED TOMATO PLANTS

ITAMAR GLAZER[1], AKIVA APELBAUM[2] AND DANIEL ORION[1]
[1]DEPT. NEMATOLOGY, [2]DEPT. FRUIT AND VEGETABLE STORAGE,
ARO, THE VOLCANI CENTER, BET DAGAN, ISRAEL

Tomato plants infected with the root-knot nematode (RKN) <u>M. javanica</u> produce ethylene at a rate several fold higher than the uninfected plants (1). This increase coincides with an increase in 1-amino cyclopropane-1-carboxylic acid (ACC) level in the plant's root and leaves (2). When excised roots were treated with aminoetoxyvinylglicine (AVG) or aminooxyacetic acid the nematode-induced ethylene production was inhibited, indicating that the nematode-infection induced ethylene production by accelerating the rate of ACC formation from S-adenosyl-methionine. Altering the rate of ethylene production in the infected roots did not affect nematode development up to the 3rd stage of juveniles, nor did it affect nematode penetration and initiation of gall formation. However, development of adult nematodes was markedly inhibited by both ethylene stimulators or inhibitors, whereas the rate of gall growth was accelerated by stimulators of ethylene production and suppressed when the production or action of the hormone was inhibited. Examination of fractured galls with SEM revealed that the hypertrophied cortical parenchyma of the galls treated with AVG, was less developed than the untreated control. On the other hand, the parenchyma of galls treated with supraoptimal concentration of auxin was larger in diameter than that of the untreated galls, due to the induction of ethylene production by the auxine. In all treatments the vascular cylinder was not affected in size and structure, nor were the coencytes induced by the RKN <u>M. javanica</u>.

Nematode infection inflicted growth inhibition in tomato plants, about 40% inhibition in shoot fresh weight was observed on the 30th day after inoculation. Foliar spray with inhibitors of ethylene production or action resulted in a recovery from the inhibition caused by the nematode infection. Foliar spray with 10 µM AVG applied to nematode-infected plants resulted in a partial recovery from the growth inhibition

Y. Fuchs and E. Chalutz (eds.) Ethylene: Biochemical, Physiological and Applied Aspects.
ISBN 90-247-2984-X. Printed in The Netherlands
©1984, Martinus Nijhoff/Dr W. Junk Publishers, The Hague.

abscission are so precisely localized to these cells. It has been proposed that cells of abscission zones constitute a determinant class of positionally differentiated ethylene responsive target cells (Type II) which are distinct in function and performance from their neighbour cells [5,6]. In this presentation to the Lieberman Symposium the evidence will be briefly summarised for the view that abscission is necessarily dependent upon the prior differentiation of cells of a determinant class with ethylene-specific biochemical responses. New experimental information will then be presented in support of the abscission target cell concept. Lastly, the question is raised as to whether the abscission of all organs in plants is dependent upon the differentiation of ethylene responsive target cells. Evidence will be presented that fruits of the Gramineae possess abscisic acid target cells

Target Cells

From light microscopy and scanning and transmission electron microscopy it appears that the ethylene target cells normally constitute an almost continuous sheet at the interface between the potentially shedding organ and permanent structural tissues of the plants, eg. at the base of leaf petioles, petals and sepals or at the base of bud, flower or fruit pedicels. It is clear that not every cell is necessarily a target cell: epidermal cells rupture rather than separate and in flower buds of Ecballium elaterium discrete groups of target cells are differentiated across the presumptive abscission region, so that pockets of separated cells are produced like the perforations in a tear-off strip of paper [7]

Many studies of the novel enzyme activities induced during abscission have shown them to be restricted to the cells of the abscission zone. The original experiments, with 1mm slices of the pulvinus-petiole abscission zone of Phaseolus vulgaris, showed the induction by ethylene (and the repression by auxin) of a B-1:4-glucanase (cellulase) activity in those cells at the distal border of the petiole tissue (the dome) at the junction with the pulvinus [8]. This induction did not occur in non-dome tissue above or below the zone. Dome tissue (300 um in thickness) is also the region of enhanced protein and nucleic acid synthesis in response to ethylene (repressed by auxin) as shown by autoradiography. This region approximates to the somewhat irregular sheet of one to two rows of cells in which expansion growth is initiated by ethylene (but repressed by auxin) concurrently with the induction of the B-1:4-glucanase activity. Both events precede the separation of the dome and pulvinus cells at abscission [9].

Since these early experiments, the ethylene induced (and auxin suppressed) de novo synthesis of a novel abscission isozyme of B 1:4-glucanase (pI 9.5) as distinct from the constitutive and auxin enhanced acidic isozyme (pI 4.5) has been demonstrated in vivo [10,11]. More recently ethylene has been shown to accelerate (and auxin repress) the accumulation of the mRNA for abscission (pI 9.5) cellulase. Translation of this mRNA in in vitro cell free systems has shown the product to co-precipitate with antibodies to purified abscission cellulase though the protein is of lower molecular weight (42,000 daltons compared with 52,000). Since active secreted cellulase appears some hours after the mRNA production, post-translational glycosylation may be a pre-secretion requirement. That this enzyme is concerned either with wall loosening or with middle lamella degradation seems evident from the suppression of abscission events that results when the zone region is treated with antiserum (but not control serum) to abscission cellulase antigen.

In the wealth of abscission literature polygalacturanases, 1:3-glucanases, pectinases, chitinase, peroxidases, nucleases and proteases have all been implicated and may all have a part to play in abscission [2]. Unlike pI 9.5 cellulase however, none of the other enzyme changes are known to be confined to the zone itself. Also, since such a small proportion of the cells that are sampled and extracted qualify as ethylene target cells one must question how many of these enzyme changes reflect the response of target cells and not those of their close non-target neighbours.

Electron microscope studies with the rachis abscission zone of Sambucus nigra, have shown that those cells that are induced to separate from each other by ethylene exhibit ultrastructural changes including enlargement of dictyosome stacks, the production of numerous secretory vesicles and the considerable proliferation of rough endoplasmic reticulum. These ultrastructural changes, which are highly suggestive of induced protein synthesis and secretory activity are not seen in adjacent non-separating cells, nor in zone cells in which separation is repressed by auxin [12].

Although secretory activity and zone cell separation are related

events the reasons for cell enlargement and middle lamella dissolution
remaining so highly localized is not known. Secreted proteins (e.g.
B-1:4-glucanase) are known to diffuse in the apoplast to cells remote
from those in which they are synthesized. None the less, only zone cells
enlarge and separate. This suggests that not only is the perception and
response of zone cells different from their neighbours but the structure
and chemical organization of their cell walls and middle lamellae are
also determined in a dissimilar way.

By analogy with animal systems, cell determination is essentially the
acquiring of a long term memory for those developmental signals which in
turn dictate the way that the cell shall respond to future signals.
Determination also implies that the memory should be maintained even when
neighbour cells are (or become) of a different determination. This means
that each cell possesses a competence (or target status) to respond in a
specific way to particular chemical or environmental signals and equally
lacks the competence to respond to other signals. Such responses are the
expression of specific molecular determinants, usually proteins which in
turn are identifiable biochemical markers of the differentiated target
state. The presence of such markers indicate a cell's competence in
signal recognition and response including the expression of new specific
determinants. In order therefore to verify target status for abscission
zone cells, it is necessary to demonstrate that they possess unique
molecular determinants before the ethylene induced response is evoked.
By further analogy with animal systems [13] an abscission target cell
once "turned on" by an inducer should be "turned off" again by withdrawal
of that inducer but retain the competence to be "turned on" again when
the inducing agent is re-introduced. Evidence towards fulfilling both
these criteria for establishing the target status of abscission zone
cells is now described.

Protein determinants in abscission cells

There is difficulty in seeking specific protein determinants in a tissue
such as P.vulgaris, where the target cells of the pulvinus-petiole layer
contribute only 1-2 rows in a dome-shaped sheet across the zone. In the
leaf rachis abscission zones of S. nigra, just distal to insertion of the
leaflets, target cells constitute some 15-30 layers and slices can be
excised in which unevoked abscission zone cells (OZ) form the major
proportion of the tissue removed. Non-target cells (MR) can be obtained

from rachis tissue midway between the leaflet (and rachis zone)
positions. Ethylene evoked zone cells (Z) can be collected as clusters
of free but intact cells when an abscission zone has separated. Using
extracted and dialysed preparations of Z, OZ and MR fractionated by
polyacrylamide gel electrophoresis in denaturing and non-denaturing
conditions unique protein bands have been resolved in Z and OZ cells,
which are not present in equivalent protein loadings of MR eg. in
polyacrylamide-SDS gels (10%) two polypeptides (M_r c. 110,000 and 95,000
daltons) are markers of Z and OZ cells.

Antisera from rats immunized with similar protein preparations of Z
and OZ has been used to recognise unique antigenic determinants in
extracts of Z, OZ and MR and leaf. Although Ouchterlony double diffusion
visualizes many determinants common to all these tissues, it also
demonstrates that Z and OZ possess common antigenic determinants not
present in MR. As might be expected, Z also contains determinants not
present in either OZ or MR. (Fig.1). Competition diffusion (intra-gel
absorption) in which Z antigen must pass an OZ antiserum barrier before
meeting the Z antiserum has confirmed the presence of unique Z precipitin
determinants which are not reduced by a threefold enhancement of OZ
antiserum (Fig. 2).

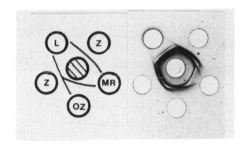

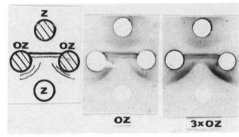

Fig. 1. Ouchterlony double immuno-
diffusion. Antigenic determinants in
S.nigra protein extracts challenged
against Z rat antiserum. [Evoked zone
(Z) unevoked zone (OZ) mid-rachis (MR)
and leaf (L)].Antiserum wells hatched.
Precipitin bands visualized with amido
black.

Fig. 2. Intra-gel absorption.
Antigenic determinants of S.nigra
evoked zone (Z) protein extracts
challenged against Z rat antiserum across an OZ or 3 x OZ
antiserum barrier. Precipitin
bands visualized with amido black

Using immuno-electrophoretic techniques, fractionated Z, OZ and MR
antigen preparations challenged with Z rat antisera IgG show distinctive
immuno-recognition arcs for each tissue. (Fig.3 a,b). However, when

electrophoretically fractionated Z, OZ and MR antigen preparations are challenged with Z antisera IgG which is first competed with MR, precipitin recognition arcs are eliminated from the MR lanes but **recognition of unique Z determinants is confirmed. (Fig.3c).**

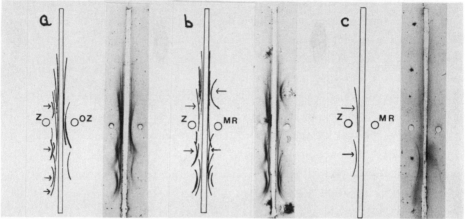

Fig. 3. (a,b) Antigenic determinants of S. nigra. Z, OZ and MR protein extracts fractionated in agarose gels and challenged against Z rat antiserum IgG. (c) Fractionated Z and MR antigens challenged against Z antiserum IgG pre-competed with MR antigen. Precipitin arcs visualized with amido black.

In P. vulgaris it has not so far been possible to distinguish by immunoelectrophoretic methods differences in antigenic determinants between unevoked target tissue of the abscission zone dome (OD) compared with the adjacent petiole tissue. Crossed immunoelectrophoresis against evoked dome (D) rabbit antiserum IgG however reveals the difference between petiole (or OD) tissue before (Fig 4a) and after (Fig 4b)

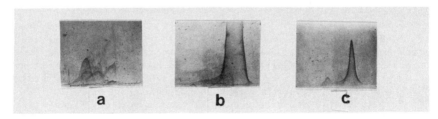

Fig. 4. Crossed immunoelectrophoresis of P.vulgaris (a) non target [Petiole (Pet)] and (b) target [Dome (D)] before and after exposure to ethylene $3-4ul.l^{-1}$ for 3 days (c) 6 days D tissue [in $3-4ul.l^{-1}$ ethylene (0-6 days) and treated with IAA $10^{-3}M$ (3-6 days] challenged against rabbit dome (D) antiserum IgG. Arcs visualized with amido black.

exposure to ethylene. This illustrates the induction of unique antigenic components by ethylene in the OD tissues, in particular, a major acidic determinant is evoked which is not present in the petiole.

Maintenance of target cell competence in P.vulgaris

During abscission, the ethylene induction of B-1:4-glucanase (cellulase) activity, the changes in intracellular ultrastructure and the expression of new antigenic determinants in the abscission zone cells are closely correlated phenomena. Critical to the target cell concept of abscission dome cells is the maintenance of determination and competence when these cells are cycled through a period of ethylene induction (turning on) followed by a period of stimulus removal (turning off) and a re-induction (turning on again) by ethylene. This sequence has been demonstrated biochemically and ultrastructurally in P.vulgaris in the following way.

The zone cells are "turned on" by exposure of explants for 3 days in ethylene at 3-4 ul.l^{-1}. Cellulase activity in the dome tissue extracts was determined by viscometry [9]. The abscission domes are then "turned off" by treating with 1ul drops of IAA (10^{-5}M or 10^{-3}M) or H$_2$O in the presence of ethylene or after treatment, enclosing them with mercuric perchlorate. Cellulase activity continues to rise in domes that are continually exposed to the ethylene stimulus alone, but withdrawal of the stimulus, or the addition of auxin arrests (and with time reduces) the level of active enzyme present in the dome tissues e.g the tissue response is progressively "turned off" (Fig. 5a and b).

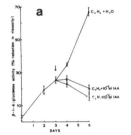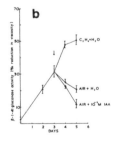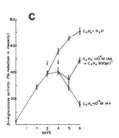

Fig. 5. B-1:4-glucanase (cellulase) activity (measured by viscometry) in dome tissue of explants induced to separate (turned "on") with ethylene 3-4 ul.l^{-1}, then turned "off" by (a) adding IAA and (b) also transferring to air and (c) turned "on" again in ethylene 500 ul.l^{-1}.

Such results are in general accord with the requirement for the continual presence of an hormonal stimulus to maintain a response in the cells of a target tissue. Regression of the evoked state can be demonstrated (Fig. 4c) by crossed immunoelectrophoretic monitoring using an antigenic preparation from auxin treated (turned off) dome tissue challenged against a similar dome antiserum IgG to that used in Fig 4a and b. The major (ethylene-induced) acidic determinant is depressed by auxin and the complex of petiole-like determinants is relatively enhanced. That dome cells retain their determinant character when turned off and can be "turned on" again even in the presence of auxin is shown by the response on exposure to high levels of ethylene (500 ul.1^{-1}). Once again, the cellulase activity increases (Fig. 5c) and ultrastructural examination of the target cells during the turning "on", "off" and "on" procedures shows a close parallel with the ethylene evoked (and auxin repressed) enzyme activity. In the "turned on" condition, electron micrographs of the 1-2 rows of target cells show active dictyosome assemblies with vesicle production and many profiles of rough endoplasmic reticulum. In the domes that remain "turned on" the endoplasmic reticulum becomes essentially smooth and the dictyosome stacks exhibit massive dilation with vesicle proliferation and shortening in length of the central non-vesiculate region. This is in marked contrast to domes from which the ethylene stimulus is removed by transfer to air and to those which auxin is supplied (turned off). In these treatments, dictyosome activity is reduced, as judged by tighter stacking and fewer vesicles. When the "turned off" dome tissue is transferred to ethylene at 500 ul.1^{-1} for two days the ultrastructural appearance of the dome cells again assumes a more "turned on" character typical of the continuously ethylene evoked target cell.

It is important to emphasize that during these experiments with P. vulgaris the ultrastructural changes we describe are observed only in the specific cells that constitute the abscission zone; they are not evinced by neighbour cells of the petiole. We believe that for the first time in plants, a successful hormonal switching system has been biochemically and ultrastructurally demonstrated. The results add further support for the target cell concept of abscission and argue in favour of the positional differentiation of cells with specific molecular determinants and particular competence for a specific hormonal recognition and response.

229

Clearly, much further work is necessary to (a) fully document the unique
characteristics and antigenic components that constitute the
differentiated abscission target cell and (b) to explain why only these
cells express the specific gene complement that permits them to separate
from their neighbours on receiving the appropriate (ethylene) signal.

Are all abscission cells target cells for ethylene?

As far as we know all dicotyledon abscission zones can be induced to
separate by ethylene. In wild type members of the Gramineae however, we
now have clear evidence that the abscission zone positionally
differentiated at the base of the fruit (or disseminule) is not a target
tissue for ethylene.[14,15] Neither ethylene (up to 1000 ul.l^{-1}, nor ACC
has any promotive effect on the cell separation process, nor does AVG or
enclosure with mercuric perchlorate delay abscission [16]. Although
treatment of the fruit or glumes with auxin will delay abscission the
only hormone found to accelerate cell separation is abscisic acid. In
this family, therefore, abscisic acid fully justifies the trivial name
bestowed upon it in 1967 [17]. The absence of ethylene-induced target
cells at the base of shedding fruits of the Gramineae is of considerable
interest. Unlike the abscinding organs of most dicotyledons, the fruits
of the Gramineae yellow and commence dessication before they are shed.
In general, loss of turgor and water deficit in leaves leads to a fall in
ethylene production which is restored only by an increase in water
potential [18].

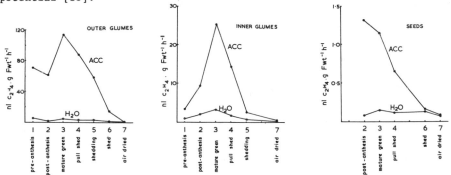

Fig. 6. Ethylene production at different stages of development by glumes
and seeds of the fruits of Avena fatua from pre-anthesis (1) to shedding
(6) and air drying (7) and their ability to convert ACC to ethylene.

Analysis of the ethylene production by glumes and seeds of Avena fatua
during fruit development together with their ability to convert ACC to

230

ethylene (Fig. 6a,b, and c) reveals a declining capacity for ethylene biosynthesis well before the abscission process is initiated. Although we have not measured abscisic acid levels in A. fatua, they have been shown to rise in maturing wheat seeds until the grains yellow which is after the water content starts to fall [19]. It seems reasonable to speculate that during evolution, plants which shed dry fruits from a well defined abscission zone like those of the wild Gramineae should have developed an alternative hormonal stimulus to one involving a biosynthetic pathway as vulnerable to desiccation as that of ethylene. It is likely therefore that abscission zones in the Gramineae have evolved positionally differentiated target cells for abscisic acid rather than target cells for ethylene.

ACKNOWLEDGEMENTS

We are indebted to Mrs Rachael Daubney and to Mrs Sheila Dunford for their expert assistance in the preparation of this manuscript.

REFERENCES
1.   Addicott FT. 1970. Biol.Rev.45: 485–524.
2.   Sexton R, Roberts JA. 1982. Ann.Rev.Pl.Physiol. 33: 133–162.
3.   Jackson MB, Osborne DJ. 1970. Nature 225: 1019–1022.
4.   Osborne DJ, Sargent JA. 1976. Planta 132: 197–204.
5.   Osborne DJ. 1977. Sci.Prog.,Oxf. 64: 53–65.
6.   Osborne DJ. 1982. Plant Growth Substances. Ed. P.F. Wareing
     Academic Press (London) pp. 279–290.
7.   Wong CH, Osborne DJ. 1978. Planta 139: 103–111.
8.   Horton RF, Osborne DJ. 1967. Nature 214: 1086–1088.
9.   Wright M, Osborne DJ. 1974. Planta 120: 163–170.
10. Lewis LN, Koehler DE. 1979. Planta 146: 1–5.
11. Koehler DE, Lewis LN, Shannon LM, Durbin ML. 1981. Phytochem.
     409–412.
12. Osborne DJ, Sargent JA. 1976. Planta 130: 203–210.
13. Hayward MA, Mitchell TA, Shapiro RJ. 1980. J.Biol.Chem. 255:
     11308–11312.
14. Sargent JA, Osborne DJ, Edwards R. 1981. XII International Botanical
     Congress. Sydney. Abstracts p.63.
15. Osborne DJ. 1983. British Plant Growth Regulators Group News
     Bulletin 6: 8–11.
16. Sargent JA, Osborne DJ, Dunford SM. 1984. J.Exp.Bot. in preparation.
17. Addicott FT, Lyon JL. 1969. Ann.Rev.Pl.Physiol. 20: 139–164
18. Wright M. 1974. Planta 120: 63–69.
19. McWha JA. 1975. J.Exp.Bot. 26: 823–827.

IS ETHYLENE THE NATURAL REGULATOR OF ABSCISSION?

PAGE W. MORGAN

Department of Soil & Crop Sciences, Texas A&M University, College Station, Texas 77843, USA

INTRODUCTION

Ethylene and abscission have been linked in the literature for many decades, but only recently has ethylene's role been supported by critical evidence. Thus it seems almost paradoxical to consider whether ethylene is the regulator of abscission. The question is made appropriate by attention to the involvement of ethylene in a natural abscission and to the mechanism(s) by which ethylene initiates abscission.

AN ABSCISSION MODEL

Description. Confusion exists about abscission due to differences in how the process has been conceptualized. Putting the process in a developmental model can help interpret the available data.

One model (9,30) recognizes three phases in the life of the expanded leaf (Fig. 1). During the leaf maintenance phase auxin prevents abscission by repressing genes for hydrolases in the abscission zone (33). Cytokinins, with their ability to delay senescence, and gibberellins, acting as growth promoters, would also prevent abscission during this phase. During the shedding induction phase, the growth-positive hormones decrease and the senescence-promoting substances such as ethylene and ABA increase. The senescence-promoting substances act by reducing the activity of the growth-promoting substances. Ethylene appears to lower auxin activity in the abscission zone by changing its synthesis, transport, destruction, or conjugation (29), making zone cells more sensitive to ethylene and more susceptible to the separation events. The leaf enters the third or shedding phase in which genes are derepressed by the decline in auxin and the increase in ethylene. The hydrolases thus synthesized cause cell separation and disruption in the abscission zone. Ethylene also promotes target-cell expansion in the abscission zone generating sheer forces which aid in the separation process (33). In

Y. Fuchs and E. Chalutz (eds.) Ethylene: Biochemical, Physiological and Applied Aspects.
ISBN 90-247-2984-X. Printed in The Netherlands
©1984, Martinus Nijhoff/Dr W. Junk Publishers, The Hague.

232

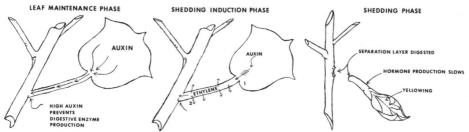

Fig. 1. The hormone balance model of leaf abscission. The sequence is from left to right. In the shedding induction phase (center), 1 indicates the action of ethylene to reduce auxin transport and to promote senescence; 2 indicates the action of ethylene and ABA to induce hydrolase synthesis. Other details in text.

addition, ethylene promotes the secretion of hydrolases (3), thereby facilitating their action to breakdown cellulose, pectins, and other cell wall materials.

Ethylene modification of auxin transport. A major evidence for the hormone balance model of abscission was the ability of ethylene to inhibit auxin transport and concomitantly promote abscission (9). Cotton cotyledons exhibited an increased level of ethylene and a decreased auxin transport capacity as they senesced. The internal levels of ethylene were high enough to reduce auxin transport and to promote abscission if applied to younger plants prior to cotyledonary abscission. Later, ethylene's ability to induce abscission and to inhibit auxin transport were shown to be parallel, and both these effects were shown to persist only so long as ethylene is present (10). The temperature sensitivity of abscission paralleled auxin transport inhibition, and both phenomena were blocked by auxin pretreatment (10). Abscission was induced by exposing the leaf blade, but not the petiole only, to ethylene; auxin transport capacity was reduced in the former but not the latter (11). Applied auxin transport inhibitors promote ethylene-mediated abscission (31); thus, auxin and ethylene levels fluctuate in agreement with the model (Fig. 1).

By its very name, the hormone balance hypothesis, this model argues that ethylene is not the regulator of abscission. Indeed, an operational dogma of modern plant physiology is that all complex developmental events are regulated by multiple regulatory substances acting in concert. However, this dogma simply requires that the question of whether ethylene is the regulator of abscission be asked in more specific terms. For example, does the leaf proceed from the maintenance phase to the shedding induction phase due to an increased level of ethylene, or in some cases is ABA the signal without ethylene?

Abscission as a whole plant phenomenon. The perspective of abscission which the model (Fig. 1) allows is one of a complex process with multiple potential control points. The occurrence of sufficient levels of auxin, cytokinins, and and gibberellins plus the energy/ metabolic components, photosynthate, salts, and water are essential for the continuation of the leaf maintenance phase. The signal to end this phase could be either a decline in the growth-positive substances, an increase in the growth-negative substances, or both.

The leaf maintenance phase also allows one to see the explant in the proper perspective. Excision immediately puts an abscission zone into the late abscission initiation phase because it a) removes leaf blade, roots and apical meristem as sources of abscission-delaying hormones, b) disrupts the water supply, c) disrupts the supply of photosynthate and salts, d) induces wound ethylene synthesis, and e) exposes tissue to possible invasion by microorganisms. Thus, while the explant does allow one to focus on the biochemical events of the abscission zone, it does not facilitate the study of the natural regulation of abscission. Yet studies with explants are often cited in ways that favor misunderstanding of the role of the growth-positive hormones during leaf maintenance phase and the transition from that phase to the abscission-induction phase.

GENERAL EVIDENCE THAT ETHYLENE REGULATES ABSCISSION

Role in senescence. Since all leaves senesce but not all abscise, senescence is considered to be a more general phenomenon than abscission. A late project of Dr. Morris Lieberman, the scientist to whom this conference is dedicated, was an investigation of the role of ethylene in leaf senescence. In this study (6) and in others (21,24) evidence indicates that ethylene regulates senescence in leaves that do not normally abscise. In addition, considerable evidence indicates that ethylene regulates senescence in flower petals (38). Thus, ethylene's role in senescence regulation seems to be rather general with abscission being a specialized effect in the more general theme.

Abscission of other organs. No major differences are known between leaf abscission and abscission of flower buds, flower parts, and fruits. Evidence reviewed in 1972 suggested that ethylene regulates the abscission of reproductive structures (1), and new studies show that ethylene signals or promotes the abscission process. Elevated ethylene production was shown to be associated with flower and fruit abscission in apple, cherry, and red raspberry (14). Production of ethylene by cotton flowers during and immediately after anthesis reached levels which would explain abscission (27). The dehiscence of some fruits is an abscission process, and extensive evidence indicates that ethylene regulates dehiscence (26). Nutritional and water stresses increase ethylene production and flower or fruit abscission (20,22). Since ethylene appears to regulate flower and fruit abscission as well as fruit dehiscence, it follows that these findings suggest a similar role in leaf abscission.

SPECIFIC EVIDENCE THAT ETHYLENE REGULATES ABSCISSION

Amounts of ethylene produced by leaves. Leaves do not produce as much ethylene as several other types of tissues; however, the issue is whether they produce enough ethylene to promote abscission. The pulvinar tissue of several species produce a peak of ethylene during senescence and prior to separation (23). Ethylene was concluded to induce separation in senescent tissues. Subsequently, in intact cotton cotyledons, both ethylene production rates and internal levels reached physiologically active levels prior to senescence (9). Thus, ethylene's role in both the shedding induction phase (9,23) and the separation phase (Fig. 1) was supported by early in the 1970's.

More recent reports indicate that leaves produce abscission-initiating levels of ethylene. As the complex leaves of Melia azedarach L. (China tree) approached autumn leaf fall (32), ethylene production rates consistently exceeded 4.5 ul $kg^{-1}h^{-1}$ at least 24 hours before leaflet abscission reached 10 percent. Ethylene concentrations in basal leaves immediately after harvest approached 1 ul $liter^{-1}$ when abscission was in progress. Hypobaric treatment blocked abscission which was restored by either termination or termination and exposure to ethylene. A climacteric-like peak of ethylene production was noted in bean leaves ten days prior to the earliest natural abscission but after chlorophyll breakdown began (6). In several other studies, leaves produced abscission-active levels of ethylene prior to separation (33,35). Space prevents listing cases where ethylene has been implicated in stress-induced abscission. Where the timing of both ethylene production and abscission have been carefully observed, ethylene levels rise sufficiently before abscission for the ethylene to initiate separation.

Diagnostic tests for ethylene involvement. Once leaves were shown to produce enough ethylene to promote abscission, other tests were employed. Reduced pressure facilitates diffusion of gases thus reducing internal levels (16); exposure to reduced pressure has been shown to delay or prevent abscission (18,26,32). CO_2 is a competitive inhibitor of ethylene action (17) and has frequently delayed or prevented abscission (see Table 9-3 in 1). Ag+ blocks ethylene action in a wide variety of responses (12) and delays or prevents abscission (25,39). AVG, at concentrations which do not inhibit net protein synthesis (28), specifically inhibits ACC synthase (15,41). AVG has been shown to inhibit or delay abscission of leaves, flowers, and fruits (8,25,37,40). Finally, sodium benzoate inhibits the conversion of ACC to ethylene (7) and also inhibits leaf abscission (25). Recently AVG, Ag+, and sodium benzoate were used in a single study and consistently showed that ethylene regulates bean explant abscission (25). Thus the varied diagnostic tests for ethylene involvement (removal, competitive inhibitor, inhibitor of action and inhibitor of synthesis) all prevent or delay abscission.

The chance that any of these tests involve an artifact, that they effect some abscission-dependent process which does not respond to ethylene, is greatly diminished by their relatively large number and agreement, the differences in how they influence ethylene activity, and their demonstrated degree of specificity.

Other evidence for ethylene involvement. Does ethylene initiate abscission or only promote it? Inhibition of abscission by actinomycin D, determinations of RNA and protein synthesis, and monitoring of cellulase synthesis led to the conclusion that ethylene induces de novo synthesis of hydrolases active in abscission (1,2). This conclusions then argues that ethylene must do more than promote abscission.

A narrow row of "target cells" in the abscission zone enlarge in response to ethylene but not auxin (33), while cells in the petiole elongate in response to auxin but not ethylene. The target cells accumulate and discharge cytoplasmic vesicles into the cell wall; these changes, along with the synthesis of cellulase, are promoted by ethylene and prevented by auxin. The unique responses of target cells to ethylene, the opposite effects of auxin, and the consistency of their behavior with a role in the regulation of abscission are powerful arguments that both auxin and ethylene have specific roles in the natural regulation of abscission. What is not yet known is the responses of these cells to ABA, cytokinins, giberrellins, and related compounds.

Aseptic tissue metabolizes ethylene, and this metabolism is correlated with physiological responses to ethylene (13). Two inhibitors of ethylene action, CO_2 and Ag^+, are presumed to block the active site, and both block abscission (1,12). An increase in the oxidation of ^{14}C-ethylene to $^{14}CO_2$, in response to abscission-promoting removal of the leaf blade, in the abscission zone but not in adjacent petiole tissue, accompanies cotton leaf abscission (13). The effect of leaf blade removal is reversed by auxin. The correlation of ethylene metabolism and abscission currently stands as additional evidence that ethylene has a specific hormonal role in the abscission zone.

EVIDENCE THAT ETHYLENE DOES NOT REGULATE ABSCISSION

ABA as an abscission regulator. Recently the information that ABA regulates abscission has been reviewed (4,5). The evidence is summarized as follows: a) the presence of ABA in leaves, branches, flowers, young fruit, and mature fruit is correlated with the occurrence of abscission; b) application of ABA promotes abscission; and c) ABA will cause abscission in the absence of ethylene. The association of ABA with senescence is extensive; however, since ABA promotes ethylene synthesis, a specific role for ABA in abscission regulation must be questioned. It should be recognized that the correlation between ABA occurrence and abscission (5) does not establish a role. Actually, ABA may influence general cellular senescence and make tissue more sensitive to its own ethylene. As yet there is no evidence that ABA specifically moves to or accumulates to senescence/abscission promoting levels in the abscission zone. In all cases where applied ABA promotes abscission, the effect may be due to either stimulation of ethylene production, nonspecific effects on senescence, or effects on abscission independent of ethylene.

Does ABA promote abscission via an effect on ethylene production? Since ABA promotes ethylene synthesis, it may promote abscission indirectly (5,37). ABA promotes citrus leaf explants to abscise and also stimulates ethylene production (37). When AVG was applied, both ABA-mediated abscission and ethylene production were reduced. Subsequently, ethylene applied to the explants treated with both ABA and AVG restored abscission. It was concluded that ABA induced ethylene production and the ethylene induced abscission. Subsequently, both AVG and Ag+ were used to evaluate the ability of ABA to promote the senescence in carnation flowers (36). ABA promoted ethylene production. Ag+ did not influence the peak of ethylene production, but it prevented the appearance of visual senescence symptoms. AVG delayed and reduced the peak of ethylene production and delayed the development of sensescence. ABA was concluded to promote senescence via its stimulation of ethylene production.

In these studies (36,37), AVG was used in the concentration range (0.1 mM) which had previously been shown to inhibit ethylene production without reducing protein synthesis (28). The specificity of the AVG effect in the leaf explant study (37) was further verified by restoring abscission in the ABA plus AVG treated explants with ethylene. If the inhibition of ABA-induced abscission is due to an artifactual effect of AVG other than the inhibition of ACC synthase, it must be an effect that does not influence the action of ethylene in abscission. The capacity of Ag+ to block the promotion of flower senescence by ABA (36) argues that the effect of AVG in these studies is not an artifact. The current evidence indicates that the major action of ABA in abscission is indirect, through its effect on ethylene production.

Abscission in the absence of ethylene. Arguing that ABA is the abscission regulator, cases have been put forward where abscission, occasionally ABA-induced, occurred in the "absence of ethylene" (4,5). One case involved the use of vacuum to remove ethylene (18), the least satisfactory test for ethylene involvement because the disruption is attempted after ethylene is formed. Further, ethylene accumulates in orange fruits (37) under vacuum conditions similar to those used in the studies cited (18). ABA promotion of abscission was only partially removed by 10 percent CO_2 (19), but again CO_2 does not always completely block ethylene action (1,37). Another study is cited in which young cotton explants abscised in an enclosed flowering system without producing a preseparation peak of ethylene (5). The question here is not whether a peak of ethylene was produced but whether enough ethylene was produced to activate the events in the target cells. The latter condition was possibly achieved without being detected or excluded by the methods employed (5).

In the alleged cases of abscission in the absence of ethylene an alternate explanation is possible. First, ethylene may have been present and acting at a threshold level to specifically activate hydrolase synthesis in target cells. Secondly, ABA may have had an effect independent of ethylene which promoted abscission. For example, ABA may have hastened membrane deterioration, promoting hydrolase secretion into the intercellular spaces (see 5). This might also partially explain ABA's modest promotion of abscission and cellulase activity in saturating levels of ethylene (19). If both hydrolase synthesis and hydrolase secretion are limiting steps in abscission, ethylene and ABA could influence both or either in ways that could produce the results under discussion (5). Abscission in the absence of ethylene is not an established fact.

HORMONE BALANCE AND ABSCISSION

Auxin and ethylene. Study of the seasonal progression of ethylene production and auxin levels prior to abscission (35) revealed that these factors were positively correlated in young tissues and were not correlated in mature tissues approaching abscission. In leaves of two species, auxin levels fell and ethylene levels remained the same or increased; in leaves and fruit of a third species, ethylene production increased and auxin levels remained constant. Two conclusions are apparent: a) ethylene production is not controlled by auxin in tissue progressing into senescence; b) the relative balance of auxin and ethylene can regulate abscission. The latter conclusion agrees with a theme common throughout the abscission and fruit-ripening literature (1,34), namely that sensitivity to ethylene as well as the absolute level of ethylene is important in determining the tissue response to the gas. The same may be true for ABA. This emphasizes the importance of the total hormone complement rather than the action of a single substance.

Conclusion. The abscission process represents the typical scientific enigma. The more that is learned, the more that is not known. At this point it seems possible that ethylene may be a universal, specific signal for abscission. On the other hand, it seems very unlikely that it ever acts under natural conditions without the involvement of auxin, ABA, and various growth-promoting substances. Thus, it does not appear to be the exclusive signal for abscission.

REFERENCES

1. Abeles FB. 1973. Ethylene in plant biology. Academic Press, NY
2. Abeles FB. 1968. Role of RNA and protein synthesis in abscission. Plant Physiol 43: 1557-1586 (and references therein).

3. Abeles FB, Leather GR. 1971. Abscission: Control of cellulase secretion by ethylene. Planta 97: 87-91.
4. Addicott FT. 1982. Abscission. University of California Press, Berkeley.
5. Addicott FT. 1983. Abscisic acid in abscission. In Addicott, FT ed, Abscisic Acid. Praeger Publishers, New York. pp 269-300.
6. Aharoni N, Lieberman M and Sisler L. 1979. Patterns of ethylene production in senescing leaves. Plant Physiol 65: 796-800 (and references therein).
7. Apelbaum A, Wang SY, Burgoon AC, Baker JE, Lieberman M. 1981. Inhibition of the conversion of 1-amino-cyclopropane-1-carboxylic acid to ethylene. Plant Physiol 67: 74-79.
8. Bangerty F. 1978. The effect of substituted amino-acid on ethylene biosynthesis, respiration, ripening and preharvest drop of apple fruits. J Amer Soc Hortic Sci 103: 401-4.
9. Beyer EM Jr and Morgan PW. 1971. Abscission: The role of ethylene modification of auxin transport. Plant Physiol 48: 208-212.
10. Beyer EM Jr. 1973. Abscission: Support for a role of ethylene modification of auxin transport. Plant Physiol 52:1-5.
11. Beyer EM Jr. 1975. Abscission The initial effect of ethylene is in the leaf blade. Plant Physiol. 55: 322-327.
12. Beyer EM Jr. 1976. A potent inhibitor of ethylene action in plants. Plant Physiol 64: 971-974.
13. Beyer EM Jr. 1979. [^{14}C] ethylene metabolism during leaf abscission in cotton. Plant Physiol 64: 971-974 (and references therein).
14. Blanspied GD. 1972. A study of ethylene in apple, red raspberry, and cherry. Plant Physiol 49: 627-630.
15. Boller T, Herner RC, Kende H. 1979. Assay for and enzymatic formation of an ethylene precursor, 1-aminocyclopropane-1-carboxylic acid. Planta 145: 293-303.
16. Burg SP, Burg EA. 1966. Fruit storage at subatmospheric pressures. Science 153: 314-315.
17. Burg SP, Burg EA. 1967. Molecular requirements for the biological activity of ethylene. Plant Physiol 42: 144-152.
18. Cooper WC, Horanic G. 1973. Induction of abscission at hypobaric pressures. Plant Physiol 51: 1002-1004.
19. Cracker LE, Abeles FB. 1969. Abscission: Role of abscisic acid. Plant Physiol 44: 1144-1149.
20. El-Beltagy AS, Hall MA. 1974. Effect of water stress upon endogenous ethylene levels in Vicia faba. New Phytol 73: 47-60.
21. Gepstein S, Thimann KV. 1981. The role of ethylene in the senescence of oat leaves. Plant Physiol 68: 349-354.
22. Guinn G. 1982. Fruit age and changes in abscisic acid content, ethylene production, and abscission rate of cotton fruits. Plant Physiol 69: 345-352 (and references therein).
23. Jackson MB, Osborne DJ. 1970. Ethylene, the natural regulator of leaf abscission. Nature 225: 1019-1022.
24. Kao CH, Yang SF. 1983. Role of ethylene in the senescence of detached rice leaves. Plant Physiol 73: 881-885.
25. Kushad MM, Poovaiah BW. 1984. Deferral of senescence and abscission by chemical inhibition of ethylene synthesis and action in bean explants. Plant Physiol. In Press.
26. Lipe JA, Morgan PW 1972. Ethylene: Response of fruit dehiscence to CO_2 and reduced pressure. Plant Physiol 50: 765-768 (and references therein).

27. Lipe JA, Morgan PW. 1973. Ethylene, a regulator of young fruit abscission. Plant Physiol 51: 949-953 (and references therein).

28. Mattoo AK, Anderson JD, Chalutz E, Lieberman M. 1979. Influence of enol ether amino acids, inhibitors of ethylene biosynthesis, on aminoacyl transfer RNA synthetases and protein synthesis. Plant Physiol 64: 289-292.

29. Morgan PW. 1976. Ethylene physiology. In: Audus LJ, ed, Herbicides: Physiology, Biochemistry, Ecology. 2nd Edition, Volume 1: 255-280, Academic Press, London.

30. Morgan PW. 1984. Chemical manipulation of abscission and desiccation. In Hilton JL, ed., Agricultural Chemicals of the Future (BARC Symposium 8). Rowman and Allanheld, Publishers. In Press.

31. Morgan PW, Durham JI. 1972. Abscission: Potentiating action of auxin transport inhibitors. Plant Physiol 50: 313-318.

32. Morgan PW, Durham JI. 1980. Ethylene reduction and leaflet abscission in Melia azedarach L. Plant Physiol. 66: 88-92.

33. Osborne DJ, Sargent JA. 1976. The positional differentiation of abscission zones during the development of leaves of Sambucus nigra and the response of the cells to auxin and ethylene. Planta 132: 197-204.

34. Pratt HK, Goeschl JD. 1969. Physiological roles of ethylene in plants. Ann Rev Plant Physiol 20: 541-584.

35. Roberts JA, Osborne, DJ. 1981. Auxin and the control of ethylene production during the development and senescence of leaves and fruits. J Exp Bot 32: 875-887.

36. Ronen M, Mayak S. 1981. Interrelationship between abscisic acid and ethylene in the control of senescence processes in carnation flowers. J Exp Bot 32: 759-765.

37. Sagee O, Goren R, Riov J. 1980. Abscission of citrus leaf explants. Interrelationships of abscisic acid, ethylene, and hydrolytic enzymes. Plant Physiol 66: 750-753.

38. Suttle JC, Kende H. Ethylene and senescence in petals of Tradescantia. Plant Physiol 62: 267-271 (and references therein).

39. van Meetren V, de Proft M. 1982. Inhibition of flower bud abscission and ethylene evolution by light and silver thiosulphate in Lilium. Physiol Plantarum 56: 236-240.

40. Williams MW. 1980. Relation of fruit firmness and increase in vegetative growth and fruit set of apples with aminovinyl glycine. HortSci 15: 76-77.

41. Yu YB, Adams DO, Yang SF. 1979. 1-aminocyclopropane carboxylate synthase, a key enzyme in ethylene biosynthesis. Arch Biochem Biophys 198: 280-286.

ANTOMICAL ASPECTS OF CITRUS ABSCISSION - EFFECTS OF ETHYLENE ON LEAF
AND FRUIT EXPLANTS.

GOREN, R., M. HUBERMAN and E. ZAMSKI.
Faculty of Agriculture, Hebrew University of Jerusalem, Rehovot 76100, Israel

ABSCISSION - GENERAL ANATOMICAL INTRODUCTION

The anatomy of abscission has been attracting the attention of
researchers in plant science for many years (2), and Hodgon's pioneer
work on the anatomy of leaf abscission (10) in citrus was published as
early as 1918. In citrus we can define five abscission zones (AZs):
the basal and laminar AZs of leaves, the shoot/peduncle (AZ-A) and
calyx (AZ-C) AZs of fruit, and the AZ between style and ovary.

In 1948 Scott et al. (16) published the results of their study on
the basal citrus leaf abscission; later on the major effort focused on
the anatomy of the calyx abscission zone (AZ-C) of the fruit mainly at
the mature stage (9,15,18). The AZ consists of about 10 to 20 cell
layers which differ in shape and size from neighbouring cells. Nevertheless, no clear abscission layer can be detected by light microscopy
(16). During the course of abscission, the separation layer of the leaf
can be distinguished by its cells - 1 or 2 cell layers - which have
strong affinity to polysaccharides dyes (16,18). The AZ is made up almost
entirely of living parenchyma cells traversed by tracheary elements, and
its cell-walls are thin with little, if any, lignin and suberin (16).
Prior ot abscission production of tanins, deposition of suberin and
lignin (16) and accumulation of starch grains increase in AZ cells (18),
but no such changes can be observed in cells adjoining the separation
layers. The amount of calcium oxalate in AZ cells increases to a level

Y. Fuchs and E. Chalutz (eds.) Ethylene: Biochemical, Physiological and Applied Aspects.
ISBN 90-247-2984-X. Printed in The Netherlands
©1984, Martinus Nijhoff/Dr W. Junk Publishers, The Hague.

similar to that of starch grain. During the separation process, cells at the proximal side of the leaf AZ become larger (16). Recently it was reported that cells around the notch at the leaf AZ triple their size during the stages of the process (13). However, no meristematic activity was found in either leaf or fruit AZs (13,16,18). Hodgson (10) observed that abscission in citrus leaves is associated with cell-wall swelling and with gelatinization of cell-wall material which precede the separation process. Swelling of cell-walls at the AZ was later observed in AZ of Valencia fruits (3), and in stylar AZ of Citron (5). Prior to cell separation the amount of endoplasmic reticulum (ER) profiles and the number of Golgi bodies in the calyx AZ of lemons increased (9). Essentially similar cytoplasmic and cell-wall changes occurred in ethephon-treated explants, but the changes were faster (9). Chaudhri reported (4) that the first sign of abscission is the swelling of the middle lamella at the leaf AZ, followed by degradation of the primary cell-wall. Iwahori and Van Steveninck (9) found that in young lemon fruits only the middle lamella was dissolved in cells of the AZ between the ovary and the floral disc.

In general, AZ of the calyx also consists of small and densely organized cells which differ, as in leaves, from the neighbouring cells, and has vascular elements which are compactly packed. The colenchymatic supporting tissue, which accompanies the vascular bundles, is localized mostly at the distal side of the AZ (18). AZ-C is rich in starch grains which can be detected at the separation zone at the early stages of fruit development and reach a peak when the fruit matures (16). During fruit development, amount of starch grains which are located in the parenchymatic cells close to the vascular bundle zone - decrease (18). Wilson (18) believes that the accumulation of starch grains in cells of

AZ is connected to metabolic activities which are related to the maturation process of cells at the AZ. The process of abscission is further characterized by a complete dissolution of cell-wall components followed by cytoplasm plasmolysis and cell degradation which was observed at AZ-C during abscission (18).

Excision and incubation of young or mature citrus fruits is generally followed by an induced increase in the activity of cellulase and polygalacturonase (PG) in the peduncular abscission zone (8,11). However, while young fruitlets abscise at this AZ, no abscission can be detected in AZ-A 6 to 8 weeks after fruit setting (around mid-June) and during further stages of fruit development (7,8). Nevertheless, the activity of both enzymes in the non-abscising AZ-A showed some of the characteristics linking hydrolytic enzymes to cell-wall degradation and abscission. Several explanations were proposed to explain this unusual phenomnenon, and the following two working hypotheses were investigated: a) the assay of total enzymic activity may involve various isoenzymes, some of which are not involved in cell-wall degradation (6) and b) total extraction may involve endo-cellular enzymes (or specific isoenzymes) which do not participate in cell-wall degradation (11). None of these hypotheses were found to be true.

In the following we present findings related to the effect of ethylene on structural and ultrastructural changes in the leaf AZ and AZ-A of the fruit at different stages of fruit development.

EFFECT OF ETHYLENE ON THE LAMINAR ABSCISSION ZONE OF LEAVES

The laminar abscission zone (LA-AZ) is located between the petiole and the lamina and is characterized morphologically by a groove around it (13). In logitudinal section the groove appears as a furrow, 10-15 cells deep, into the cortex (Fig. 1). No separation layer can be seen within the abscission zone of attached leaves. Separation layer has been formed among LA-AZ cells only after leaves were artificially detached from the branch (Fig. 1).

The cortex in the LA-AZ is composed of relatively small parenchyma cells containing large starch grains (Figs 2 and 3). The inner cortical layer - the starch sheath - is also characterized by large starch grains. These cortical cells undergo dramatic changes while being transformed into AZ (see ref. 12 for experimental details). Leaf explants left as untreated control, abscised after 72 to 96 h. The first noticeable changes were observed in the cortex parenchyma close to the groove, 24 h after the beginning of the experiment. Ethylene ($30\mu l/l$) treated leaves (see ref. 11 for experimental details) followed, more or less, the same sequence of events as did control samples, but ethylene stimulated the processes, leading to separation, which started as early as 16 h after excision and the lamina was separated after 50 h. The number of cells which responded in the cortex and later composed the LA-AZ was doubled (about 40 cell-layers) in ethylene treated leaves. The anatomical changes that occurred in LA-AZ of all treated leaves were as following: at first a distinct separation layer started to appear in the outer cortical cells which are responsible for the opening of the groove. Later on, more cortical cells in the same radial plane joined the separation layer, as the process advanced inwards (Fig 1). Cells of the separation layer have a unique appearance: they possess a very dense cytoplasm with more organells than their neighbouring cells (Fig. 2). The second observed change was a process of cell-wall swelling in the separation layer and its adjacent cells, concomitantly with a gradual disappearance of cell-wall components. Later on, more cortical cells around the separation layer, and especially the cells on its distal side, became swollen until the entire AZ composed of swollen cell-walls. By using different cell-wall dyes (tuloidine blue and periodic acid - Shiff reagent) we were able to follow the dissolution of cell-wall components. At the end of the reaction, the swollen and loose cell-wall lost its polysachharides content. The middle lamella has also been dissolved, and eventually free protoplasts were observed among the separation-layer cells. These processes led to a separation of the lamina from the petiole.

An ultrustructural study revealed that in both leaf control explants and ethylene-treated explants, many large vesicles containing fibrous material were derived from the cytoplasm and penetrated into the vacuolar compartment or fused with the plasmalemma. Primary cell-walls became lamellated before abscission. Although there was no clear indication of wall degradation up to 48 h after exicision in control explants, the middle lamella appeared more deeply stained (Fig. 3), and later it dissolved and disappeared. No changes were noted in the number of chloroplasts and mitochondria but there was a slight increase in the amount of rough-ER, as previously noted (9). When abscission was almost completed, the primary walls dissolved whereas the plasmalemma and tonoplast remained intact.

EFFECT OF ETHYLENE ON AZ-A OF YOUNG FRUIT - THE ABSCISSION STAGE

The sequence of events leading to fruit separation at AZ-A and the effect of ethylene were the same as in leaf LA-AZ, but the former differ from the latter in some of its anatomical features: it was thicker and contained more cortical parenchyma cells, however as in LA-AZ, no distinct separation layer could be detected. All AZ cells, particulary those of the inner cortex, were loaded with starch grains and calcium oxalate crystals. During the separation process, the number of starch grains decreased while the amount of crystals increased. The final separation between cells at AZ-A seemed to occur at random, and not always in a radial plane from the groove inwards. In some cases separation took place above the groove, (on the distal side of the AZ), and it had always occurred in a convex line starting at the inner cortex and having its peak at the pith (Fig. 4). Cell divisions could be observed in the cortex of AZ-A of both control and treated fruits, but the amount of cell divisions was higher in ethylene-treated fruits. Anatomical changes were observed in control fruits only about 50 h after excision and the separation process occurred within a narrow zone of 5 cell layers. Unlike the slow reaction of the control, a marked reaction was observed 16 h after excision in ethylene-treated fruits which fully abscised 40 to 50 h after excision (Fig. 4). Dissolution of wall components was more pronounced in the AZ (Fig. 5), and a stronger reaction occurred at the distal side. The ultrastructural changes at AZ-A of control and ethylene-treated explants were similar to those described for LA-AZ of the leaf. However, microbodies, which were not detected at the LA-AZ of the leaves, were observed during excision at AZ-A of young fruits. They disappeared later on, both in control and in ethylene-treated explants.

EFFECT OF ETHYLENE ON AZ-A OF YOUNG FRUIT - THE TRANSITION STAGE

Six to 8 weeks after excision, the fruit abscises at AZ-C. There is, however, a transition period during which the process shifts from AZ-A to AZ-C. At this stage, a flat furrow is still obvious at AZ-A, and ethylene did not induce abscission any more, but caused AZ-A to swell (Fig. 6); separation occurred only at a cellular level in different areas of the cortex and pith. No specific separation layer was formed. The entire AZ - consisting in this case of 80 to 100 cell layers - responded to ethylene (Fig. 6), and the swelling of AZ-A resulted from extensive cell division and from cell expansion (Fig. 7). Cell division also occurred, to some extent, in control AZs while cell expansion was typical only to ethylene-treated explants. Cell division was observed mostly in the parenchyma, vascular cambium and the epidermal layer of the AZ (Fig. 7). As a result a peridem was formed as well as meristemoids of 2 to 4 small cells (see the asteric within a mother cell). Cell expansion was observed 24 h after ethylene treatment, but only in a radial plane of the inner cortex at the starch sheath zone. The cells increase in size 3 to 10 fold. Later expansion of additional cells was not restricted to the radial plane (Fig. 7). Concomitantly, cell expansion, which caused the rupture of the stele at a later stage, occurred in the pith and in the starch sheath area. Starch grains completely disappeared as a result of ethylene treatment, while some starch grains were still obvious in other areas of the AZ. According to Sexton (17) it is suggested that the disappearance of starch grains and the rise in soluble sugars increases water uptake and leads to an increase in turgor presure. Consequently, this may cause cell expansion and tearing along the middle lamella. The expansion of cells at the AZ, prior to abscission, is a well established phenomenon in citrus, however, during the transition period it occurs in groups of cells rather than across the abscission zone.

Two types of cell-wall changes were observed in the ethylene-induced swelling of AZ-A: a) separation of intact cells, usually between the outer and inner cortical parenchyma cells and also in the pith area, indicating that separation occurred along the middle lamella,

as previously reported for beans <u>Gossypium</u>, Impatiens, tobacco and
tomato (17) and b) breakdown of both the primary wall and the middle
lamella, as has also been found in bean and tobacco (17). Nevertheless,
no separation layer was formed in the swollen AZ-A (Fig. 6), unlike the
very young fruits explants.

EFFECT OF ETHYLENE ON AZ-A OF MATURE FRUIT - THE NON-ABSCISSION STAGE

Even though mature fruits did not abscise, an abscission process
did occur mainly in AZ-A of ethylene-treated fruits, but these were
limited to specific groups of cells organized in the starch sheath
region (Figs 8 - 10), so that <u>in toto</u> no abscission was expressed.
Following excision, several cell layers started to respond in AZ-A at
the starch sheat area. Upon excision, these cells were spherical and
had thin walls, but later they became irregular in shape and were poorly
stained by polysaccharide dyes. During incubation in air, and especially
during ethylene treatment, such cells reacted as in a separation layer,
although the cell layers were oriented in a longitudinal plane (Fig. 9),
rather than along a radial axis across the AZ as in the abscising
abscission zone (Figs 1 and 4). The cells followed the sequence of events
leading to cell separation: swelling of cell walls reduced sensitivity
to staining with polysaccharide dyes, cell wall dissolution and degradation and finally formation of a dissolved cavity (Figs 8 and 9) with
destroyed cell wall elements and free protoplasts in the starch sheath
area (Fig. 10). However, no more than 5 to 8 cell layers, in the radial
axis, responded with the above mentioned sequence of events, not even
after excision and ethylene treatment for 168 h. Those events were also
observed in the pith rays, particularly in pith parenchyma cells, but
less intense. Cell division could be observed at the same time in the mid

and outer bark and also in the vascular cambium. It occurred also in the epidermal layer due to ethylene treatment, and resultrd in the formation of a periderm tissue with radial cell rows (Figs 8 and 9). In explants left untreated in air, the above processes were essentially the same but took longer time and were less intense. However, no separation layer was formed in the radial plane of AZ-A, even after exposure to ethylene. When responding to ethylene, the parenchyma cells of the AZ lost most of their starch grains, which had accumulated earlier in the young cortex, pith and pith rays (Figs 4 and 5). The expanded cells of the starch sheath were completely depleted of starch grains (Figs 8 and 9). From September to January 6 to 10 months after fruit set, ethylene-treated fruit explants appeared to have, along the non-abscising AZ-A in the outer bark, a "sleeve" of unchanged cells that enclosed the dissolved area of the inner bark. The majority of the cells in this AZ remained unchanged, particularly in the stele and in the outer bark cells (Fig. 9). Therefore, no abscission could be observed on the surface of the AZ.

Electron micrographs of the same area show that 168 h after ethylene treatment most of the parenchyma cells had swollen walls and were at different stages of plasmolysis. There was also an increase in the amount of a densely stained material which was deposited, outside the plasmalemma, on the surface of the decomposed walls, and the walls became loose before separation. Occasionally large multivesicular bodies were derived from the cytoplasm and fused with the plasmalemma. In the plasmolyzed protoplasts, some tiny vesicles could be distinguished between the plasmalemma and the cell wall. The plasmalemma opposite these small vesicles was interrupted, while in other areas it remained intact and revealed no "holes". Three regions of dissolving wall material

250

could be observed in the walls of parenchyma cells in the AZ, in the middle lamella, around the branched plasmodesmata, and in the newly deposited microfibrils.

It can be concluded that separation does occur in non-abscission AZ-A, but only in certain "pockets" within the inner bark. The increase in activity of PG and cellulase in non-abscising AZ-A during incubation and ethylene treatment (8,11) is, in fact, accompanied by cell wall dissolution and degradation (Figs 8 and 9), indicating that there is no conflict between the concept that an increase in hydrolytic enzymes activity in the AZ leads to abscission (14,17). In the non-abscising AZ-A, however, this concept is true only for specific cells at AZ-A (Fig. 10) and not for all the cells in this zone. Therefore, although there is an increase in the activity of hydrolytic enzymes in the AZ, no abscission of fruits can be observed.

Acknowledgements   —   The authors are grateful to Mrs. Aviva Ben-David for her exellent technical assistance in preparing the plant material for electron microscope observation.

REFERENCES

1. Abeles FB. 1969. Abscission: Role of cellulase. Plant Physiol. 44:447–452.
2. Addicott FT. Abscission. University of California Press Berkeley. Los Angeles. London 369 pp.
3. Biggs RH. 1971. Citrus abscission. HortScience. 6:388–392.
4. Chaudhri SA. 1957. Some anatomical aspects of fruit drop in citrus. Ph.D. Dissertation, University Fla. Gainesville.
5. Goldschmidt EE., Leshem B. 1971. Style abscission in the citron (Citrus medica L.) and other citrus species: morphology, Physiology and chemical control with picloram. Amer. J. Bot. 58:14–23
6. Goren R., Huberman M. 1976. Effects of ethylene and 2,4-D on the activity of cellulase isoenzymes in abscission zones of the developing orange fruit. Physiol. Plant. 37:123–130.

7. Goren R. Teitelbaum G. Ratner A. 1973. The role of cellulase in the abscission of citrus leaves and fruits in relation to exogenous treatments with growth regulators. Acta Hortic. 34:359-362. (Symp. on Growth Regulation in Fruit Production, Long Ashton, England, 1972).

8. Greenberg J. Goren R. Riov J. 1975. The role of cellulase and polygalacturonase in abscission of young and mature Shamouti orange fruits. Physiol. Plant. 34:1-7.

9. Iwahori S. Van Steveninck RFM. 1976. Ultrastructural observation of lemon fruit abscission. Scientia Hort. 4:235-246.

10. Hodgson RW. 1918. An account of the mode of foliar abscission in citrus. University of California Publication (Bot.) 6:417-428.

11. Huberman M. Goren R. 1979. Exo-and endo-cellular cellulase and polygalacturonase in abscission zones of developing orange fruits. Physiol. Plant. 45:189-196.

12. Huberman M. Goren R. Zamski E. 1983. Anatomical aspects of hormonal regulation of abscission in citrus - the shoot-peduncle abscission zone in the non-abscising stage. Physiol. Plant. 59:445-454.

13. Jaffe MJ. Goren R. 1979. Auxin and early stages of the abscission process of citrus leaf explants. Bot. Gaz. 140:378-383.

14. Ratner A. Goren R. Monselise SP. 1969. Activity of pectin esterase and cellulase in the abscission zone of citrus leaf explants. Plant Physiol. 44:1717-1723.

15. Rogers BJ. 1971. Scanning electron microscopy of abscission - zone surfaces of valencia orange fruit. Planta. 97:358-368.

16. Scott FM. Schroeder MR. Turrell FM. 1948. Development, cell shape, suberization of internal surface, and abscission in the leaf of the valencia orange, Citrus sinensis. Bot. Gaz. 109:381-411.

17. Sexton R. Roberts JA. 1982. Cell biology of abscission. Ann. Rev. Plant Physiol. 33:133-162.

18. Wilson WC. Hendershott CH. 1968. Anatomical and histochemical studies of abscission of oranges. Proc. Ann. Soc. Hort. 92:203-210.

Plate I

Laminar abscission zone

Fig. 1.   Longitudinal section of leaf explant. 48h after excision and
          incubation in air. G=groove; SL=separation layer; Ss=starch
          sheath.
Fig. 2.   Same as Fig. 1, magnification of the separation layer.
Fig. 3.   Ultrastractural view of one cell of the separation layer
          from Fig. 2. Note : the large starch grains and the swollen
          cell wall. ch=chloroplast; st=starch grains; v=vacuole;
          w=cell wall.

Young fruit - abscising AZ-A

Fig. 4.   Longitudinal section of abscising fruit. - 50 h - of
          exposure to ethylene. dis=distal; pro=proximal; asteric
          =separation line.
Fig. 5.   Same as Fig. 4, magnification of the distal side of the AZ.
          Note : the swelling of cell walls and the color
          disappearance due to decrease in polysaccharides content in
          cells close to the separation line. Ci=inner cortex;
          Ss=starch sheath; St=starch grains; Ws=swollen cell wall.

Plate II

Young fruit - transition period

Fig. 6.   Longitudinal section of AZ-A. 72h incubation in ethylene.
          Sc=swelling cortex; Sv=swelling vascular cylinder.
Fig. 7.   Cross section at the swollen AZ-A. 72h incubation in
          ethylene. Note : cell expansion in the cortex and cell
          division in the cambium and cortex. Ca=cambium; P=phloem;
          x=xylem; Ss=starch sheath; asteric=meristemoids.

mature fruit-non-abscising AZ-A

Fig. 8.   Cross section in the center of AZ-A. 168 h incubation in
          ethylene. Note : cell expantion, dissolution and the
          beginning of cavity formation (star) in the inner bark.
          Ss=starch sheath.
Fig. 9.   Same as Fig. 8 but longitudinal section. Pe=periderm.
Fig.10.   Magnification of starch sheath area at AZ-A.96 h incubation
          in ethylene. Note : formation of a cavity and release of
          free protoplasts. Ss=starch sheath.

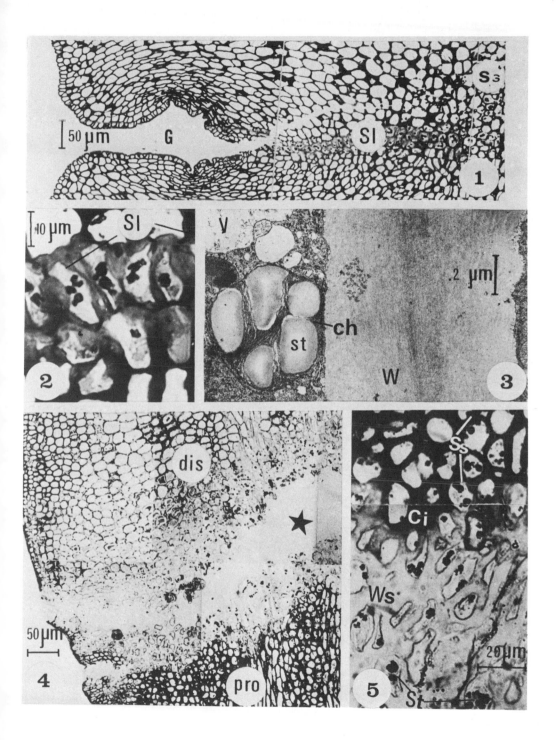

254

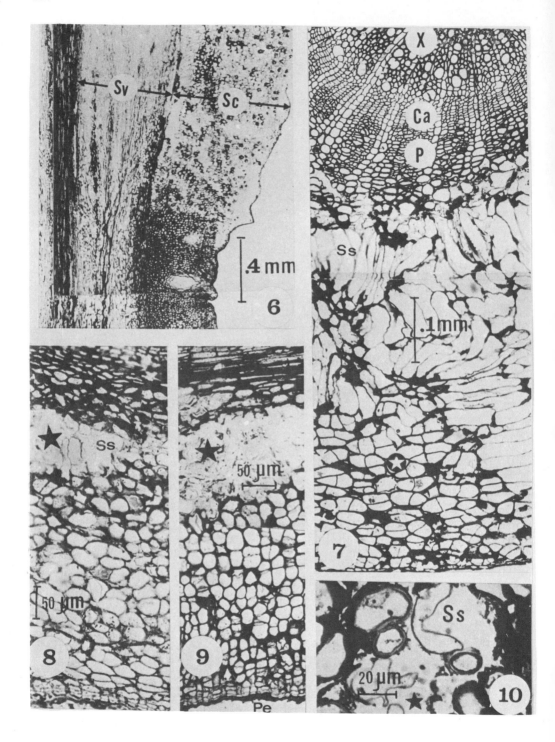

ETHEPHON ACTION IN GROWTH AND ABSCISSION CONTROL

S. LAVEE
Dep. Olei and Viticulture, Volcani Center, Bet Dagan, Israel

Ethephon is widely used in research and agriculture to produce ethylene mediated processes, such as fruit ripening, abscission and growth control.

In most cases ethephon is looked upon as a convenient form of ethylene application. The activity of ethephon is usually considered as equivalent to that of ethylene. This, however, has been shown to be only partially true. The rate of ethephon uptake, translocation, binding in the tissue and ethylene release will determine its activity. Furthermore, in some cases the physiological response to ethephon is specific and dependent not only on the rate of ethylene release, but on the structure and its molecule. The data to be discussed in this review were produced in collaboration mainly with Y. Shulman, G. Nir, Y. Ben-Tal, B. Bravdo, O. Shoseyor, E. Epstein, I. Klein, G.C. Martin and M. Kliever.

As the activity of ethephon is usually limited to the site of its application, feeding experiments with excised shoots were performed. Ethephon (ET) and Amino cyclopropane-1-carboxylic acid (ACC) were fed with the transpiration stream to excised olive shoots. The rate of ethylene release and leaf abscission along the shoots was recorded. Ethylene evolution from the lowest pair of leaves increased rapidly due to the feeding with both ET and ACC (12). However, the ethylene evolution of the ET feed shoots prevailed for a long time, while those fed with ACC were back to normal 6 hr. after the treatment regardless of application duration (Table 1).

Similar results were obtained also when increasing concentration of ET and ACC were used on a constant time basis (Fig. 1).

Leaf drop from the treated shoots occurred 50-60 hrs after their transfer to water. The leaf drop due to the natural ethylene precursor ACC was very low, while ET caused a complete leaf drop (Fig. 2).

Y. Fuchs and E. Chalutz (eds.) Ethylene: Biochemical, Physiological and Applied Aspects.
ISBN 90-247-2984-X. Printed in The Netherlands
©1984, Martinus Nijhoff/Dr W. Junk Publishers, The Hague.

Table 1: The effect of Ethephon and ACC feeding to excised olive shoots on ethylene evolution of from their leaves. (Concentration 2×10^{-3}M, ethylene ml/100 leaves/hr).

Treatment & Time of Application	Time After Transfer To Water (hr)				
	0	2	6	20	43
Control 0	22	22	26	24	19
ACC 10 - Minute	48	69	30	24	24
20	52	230	39	24	19
40	109	432	39	24	22
80	121	1426	24	24	26
160	181	1886	26	19	30
320	261	2506	17	20	27
ET 10	52	60	72	45	36
20	72	204	160	45	53
40	343	269	230	230	181
80	907	405	526	219	217
160	873	1038	814	217	167
320	897	1800	1368	1267	199

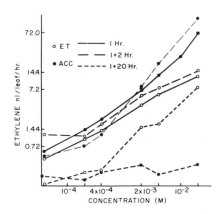

FIGURE 1:
The effect of ACC and ET concentrations on ethylene evolution from Manzanillo olive leaves following pulse feeding through cut bases of excised shoots (12).

Thus it could be concluded (12,13,14) that the rate and persistance of ethylene release determine the degree of response. A very rapid release of even high concentrations of ethylene could not lead to leaf abscission if prevailed only for a short period.

The long period of ethylene evolution from the leaves after ET treatments was shown to be due to ethylene release from the ET itself.

FIGURE 2: The effect of ACC (left) and ET (right) on leaf drop of excised olive shoots.

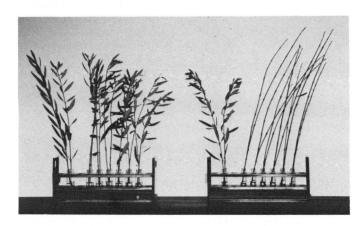

Experiments with an ethylene synthesis inhibitor AVG showed that no
enhanced ethylene synthesis was induced in the leaves due to the ET
treatments (table 2).

Table 2: The effect of AVG on ethylene evolution in ET treated and
wounded olive leaves.

Treatment	Ethylene evolution of 100 leaves ml. hr.$^{-1}$	
	−AVG	+AVG
Control leaves	2.5	2
Wounded "	21	2
Control " +ET	1230	1231
Wounded " + "	1248	1233

The rate of ethylene release from ET is both temperature and pH
dependent (3,6,7,17). At pH below 4.0 ethylene release from ET is very
small; with increase in pH the rate of release increases rapidly and
at pH 8.2 the decomposition is nearly instantly. This pH effect is,
however, greatly temperature dependent and at $20^{o}C$ ethylene evolution
was very small even at pH 7. Elevation of the temperature causes a
rapid increase in ET decomposition and at $40^{o}C$ ET at pH 7 is decomposed nearly instantly. Thus the environmental conditions and pH
of application will greatly effect the ET activity as shown in figure 3.

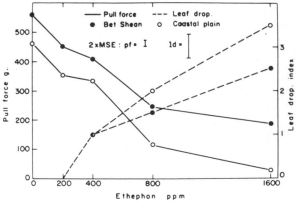

FIGURE 3: The effect of various ethephon concentrations on the fruit removal force (FRF) and leaf drop in two different climatic regions (8).

The activity of the ET on fruit and leaf abscission in the hot Bet Shean Valley was clearly lower than in the more moderate coastal plain. This is probably the result of too rapid ET decomposition in the hot region.

The response of a tissue to ET is dependent also on its physiological susceptability. Leaves on reproductive olive shoots abscised considerably more readily after an ET treatment than leaves on a similar developed vegetative shoot (Fig. 4).

FIGURE 4: Leaf drop on ET fed reproductive (left) and vegetative (right) olive shoots.

Uptake and translocation of ET other than via the transpiration stream is very limited. Studies with ^{14}C-labeled ET, showed that most of the ET applied to peach fruit, is localized in the epidermis and the adjacent cells (fig. 5).

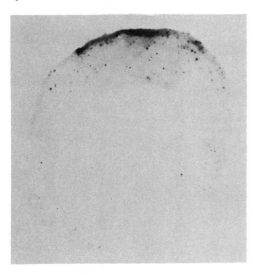

FIGURE 5: Distribution of radio-activity in ^{14}C-ET treated peach fruit. Application of 2μCi on June 23, harvested Aug. 28 (10).

With olive fruits it was shown that the fruit removal force (FRF) was reduced only when the ET was applied to the site of abscission at the proximal fruit cavity and to a lesser extent at the mid petiole (Table 3).

Table 3:	Site of application	Fruit removal force
The effect of ET application site on the fruit removal force of manzanillo olives	untreated control	628 $\pm$ 41
	proximal cavity	192 $\pm$ 27
	whole fruit	201 $\pm$ 28
	distal end of fruit	697 $\pm$ 47
(4)	pedicil	306 $\pm$ 42

The limited translocation could be demonstrated also when ^{14}C-labeled ethephon was applied to grapevine leaves. Most of the label remained after 48 hrs in the treated tissue (15). Although translocation was rather small, the site of application on the leaf had a significant effect on the degree of translocation (table 4).

Table 4: The effect of application site on ^{14}C-ET translocation in
Perlette grapevine leaves (48 hr after application ET concentration 500 mg/l).

Application Site	Total uptake %	Radio activity distribution %		
		Treated area	Rest of leaf	Petiol
upper proximal	26+3	72+3	18+3	11+2
" distal	1+2	85+3	13+2	4+1
lower proximal	21+2	72+4	2+1	20+4
" distal	19+3	90+2	5+2	5+2

The amount of ET which penetrates the epidermis is taken up rather
rapidly and can stay in the tissue as ethephon for a long time (11).

This long stability of the ethephon molecule in the tissue might be
partially due to its high affinity to conjugate with cell constituents -
particular sugars It has been shown with ^{14}C-labeled ET that the
sucrose-ethephon conjugate is formed instantly when ET is applied
(fig. 6).

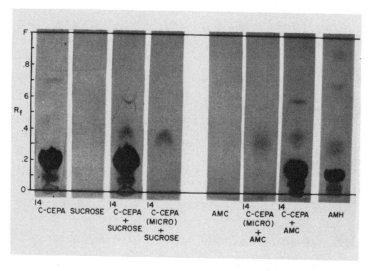

FIGURE 6: Autoradiogram of TLC chromatograms of ^{14}C-ET with and without
sucrose, and of ^{14}C-ET with and without methanol extracts of
peach fruits (AMC-acid methanol extract of non-radioactive
and AMH of radioactive ET treated fruits) (11).

These ethephon conjugates which are stable and might release the
ethephon, very slowly can account for late ethylene responses notable
after ET treatments. Leaves of olive trees, treated with ET, show first
a rapid ethylene evolution which decreases rapidly for 1-2 days, but
after 2-3 days the ethylene evolution from the leaves increases again.
It was suggested (1,2) that the late and slow increase of ethylene
evolution is involved in the induction of leaf abscission. When ET is
used at pH-7, this second increase in ethylene evolution does not occur
and leaf drop is also greatly reduced. The second increase will not
take place when ET is applied to dead leaves, although the evolving
ethylene originates from the ET only.

Bivalent ions have a marked effect on ethylene evolution from
ethephon treated olive trees. Leaf drop and FRF did not change in the
same order (Table 5).

Table 5: The effect of bivalent ions on ethylene evolution FRF and leaf
drop from ethephon treated Manzanillo olive trees (Ions
applied as 0.1M acetate salts with 1000 mg/l ethephon).

Treatment		FRF	Leaf drop	Ethylene Evolution $ml.g^{-1}.h^{-1}$	
		g.	0-5	Leaf	Fruit
control		531	0	0.4	0.3
Ethephon		205	3	12.5	30.0
"	+ Ca	350	1.7	18.6	36.1
"	+ Zn	580	1.7	3.0	6.1
"	+ Mn	480	1.7	13.1	15.2
"	+ Mg	420	2.3	16.7	18.4
"	+ Cu	380	2.0	42.0	37.8

All ions decreased the effect of ethephon of FRF reduction to a
greater or lesser extent. Ethylene evolution, however, was greatly
enhanced by some, such as Cu^{++} and markedly reduced by others, such as
Zn^{++}. These effects were not due to pH differences as that was equally
adjusted to pH-6.7. The reason for this effect is not yet clear, but it
is obvious that the ion effect on the rate of ethylene release from
ethephon cannot be the major answer for the different response to ET
of the leaves and fruits.

This could be further demonstrated when increasing concentrations of Ca-acetate were applied together with the ethephon. While the increase in Ca^{++} concentration inhibits the ethephon effect on FRF reduction, the ethylene evolution is inhibited by low concentrations and enhanced by the higher ones (table 6).

Treatment Ca-Acetate mM.	FRF g	Ethylene evolution $ml.g^{-1}.h^{-1}$
Ethephon	245	18.3
0.05+ET	235	18.0
0.1 + "	336	11.5
2.5 + "	390	11.8
5.0 + "	415	29.4
10.0+ "	541	25.9
- + - (control)	587	0.6

Table 6:
The effect of increasing Ca-acetate concentrations on ethephon induced FRF reduction and ethylene evolution from olive leaves.

This ion effect seems to be specific to ethephon. When Ca^{++} is added to another ethylene releasing compound Alsol, no effect on FRF was noted (table 7).

Treatment	FRF -gram	
	-CaAC	+CaAc 10mM
control	568	559
Ethephon	271	515
Alsol	252	258

Table 7:
The effect of Ca-acetate on FRF reduction, induced by ethephon and Alsol.

The specificity of ethephon to induce growth responses was further demonstrated on grapevine shoot growth control. Application of ethephon to growing grapevine shoots causes a marked reduction or cessation of growth of the shoot tip and an inhibition of sprouting of the lateral buds (9,18). This, however, was not the case when the Alsol was used (Fig. 7).

The ethephon induced inhibition of bud growth is due to degradation of the active tissue in the growing point dome and leaf primordia (Fig. 8).

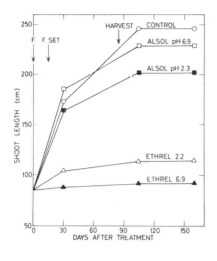

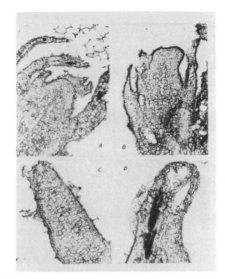

FIGURE 7: The effect of ethephon and Alsol at pH 2.3 and 6.9 on the growth of "Perlette" grapevine shoots (5).

FIGURE 8: Longitudinal section of apecis of "Muscat Hamburg" grapevine shoots. 7 days after Ethephon treatment. A. Untreated apex. B. Treated apex x75. C. Primordium of an untreated apex. D. Primordium of leaf of treated apex x188.

The ethephon effect on growth is accompanied by a reduction of the photosynthetic rate of the treated leaves. Shoseyov (16) has shown that the degree of photosynthesis reduction was concentration dependent and prevailed for long periods - at least 9 days (Fig. 9).

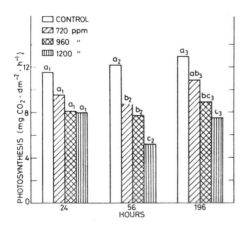

FIGURE 9:
The effect of ethephon on the rate of photosynthesis of 'Dabouki' grapevine leaves up to 9 days after treatment (different letters indicate significance of the 5% level) (16).

264

The effect of ethephon on the translocation of metabolits was also
tested by feeding the 6th leaf below the apex of a fruit bearing grapevine shoot with $^{14}CO_2$ (16). While in control shoots the label moved
mainly to the apex and in topped shoots to the upmost lateral bud, in
ethephon treated shoots in which the growth and the lateral buds were
inhibited, the label translocated mainly to the developing bunch
(Fig. 10). This change in translocation of the ^{14}C-labeled metabolits
was due to the inhibition of the more potant sinks of the developing
shoot.

FIGURE 10: The effect of ethephon and topping on the distribution of
^{14}C-labeled metabolits from leaf six to other parts of the
shoot. A. Control shoot. B. Topped shoot. C. Topped + 750
mg/l ethephon. (Application 7 days after fruit set, 2 hr
exposure, sampled 72 hr after treatment.)

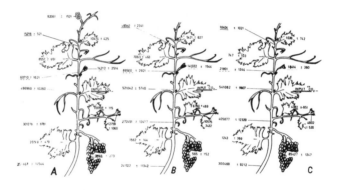

Thus, although the physiological responses induced by ethephon
treatments are mainly ethylene mediated effects, there seems to be a
specificity of the ethephon effect dependent on the conditions and plant
material. It should be concluded that ethephon has to be considered as
an active regulator having its own nature which probably acts in most
cases via ethylene. This calls for a critical consideration of the
results when ethephon is used as a "liquid ethylene" for studying
ethylene effects as such.

REFERENCES

1. Ben-Tal, Y. and Lavee, S. 1976. Increasing the effectiveness of ethephon for olive harvesting. HortScience 11: 489-490.

2. Ben-Tal, Y. and Lavee, S. 1976. Ethylene influence on leaf and fruit detachment in 'Manzanillo' olive trees. Scientia Hortic. 4: 337-344.

3. Ben-Tal, Y., I. Klein and S. Lavee. 1979. The role of the source of ethylene on the development of an abscission layer in olive pedicels. In: H. Geissbuhler, ed, Advances in Pesticide Science, Pergamon Press, New York, pp 347-350.

4. Epstein, E., Klein, I. and Lavee, S. 1977. The fate of $1,2^{14}$C-chloroethyl-phosphonic acid (ethephon) in olive. Physiol. Plant 39: 33-37.

5. Hirschfeld, G. and Lavee, S. 1980. Control of vegetative growth of grapevine shoots by ethylene-releasing substances conditions and sites of action. Vitis 19: 308-316.

6. Klein, I., Epstein, E., Lavee, S. and Ben-Tal, Y. 1978. Environmental factors affecting ethephon in olive. Sci. Hortic. 9: 21-30.

7. Klein, I., Lavee, S. and Ben-Tal, Y. 1979. Effect of water vapour pressure on thermal decomposition of 2-chloroethyl-phosphonic acid. Plant Physiol. 63: 474-477.

8. Lavee, S. 1976. Abscission studies of olive fruit - Physiological and Horticultural aspects. Olea 3: 35-56.

9. Lavee, S., Erez, A. and Shulman, Y. 1977. Control of vegetative growth of grapevine (Vitis vinifera L.) with chloroethylphosphonic acid (ethephon) and other growth inhibitors. Vitis, 16: 89-96.

10. Lavee, S. and Martin, G.C. 1974. Ethephon ($1,2-^{14}$C(2-chloroethyl) phosphonic acid) in peach fruits. I. Penetration and persistance. J. Am. Soc. Hortic. Sci. 99: 97-9.

11. Lavee, S. and G.C. Martin. 1975. Ethephon ($1,2-^{14}$C(2-chloroethyl) phosphonic acid) in peach (Prunus persica L.) fruits. III. Stability and Persistance. J. Am. Soc. Hort. Sci. 100: 28-31.

12. Lavee, S. and G.C. Martin. 1981. Ethylene evolution following treatment with 1-aminocyclopropan-1-carboxylic acid and ethephon in an in vitro olive shoot system in relation to leaf abscission. Plant Physiol. 67: 1204-1207.

13. Lavee S., and G.C. Martin. 1981. In vitro studies on ethephon - induced abscission in olive. I. The effect of application period and concentration on uptake ethylene evolution and leaf abscission. J. Amer. Soc. Hoet. Sci. 101: 14-18.

14. Lavee, S. and G.C. Martin. 1981. In vitro studies on ethylene induced olive abscission. II. The effect on ethylene evolution and abscission of various organs. J. Am. Soc. Hort. Sci. 106: 19-26.

15. Nir, G. and S. Lavee. 1981. Persistence uptake and translocation of ^{14}C ethephon (2-chloroethyl phosphonic acid) in 'perlette' and 'cardinal' grapevines. Aust. J. Plant.Physiol. 8: 57-63.

16. Shoseyer, O. 1983. Out of season grape production of one year old cuttings. M.Sc. Thesis, Hebrew University of Jerusalem, Faculty of Agriculture, Rehovot.

17. Shulman, Y., Avidan, B., Ben-Tal, Y. and Lavee, S. 1982. Sodium bi-carbonate, a useful agent for pH adjustment of ethephon controlling grapevine shoot growth and loosening olive fruits. Riv. Ortoflorro frutt. It. 66: 181-187.

18. Shulman, Y., G. Hirschfeld, and S. Lavee. 1980. Vegetative growth control of six grapevine cultivars by spray application of 2-chloroethyl phosphonic acid (ethephon). Amer. J. Enol. Vitic. 31: 288-293.

EFFECT OF ETHYLENE ON INDOLE-3-ACETIC ACID TRANSPORT, METABOLISM, AND
LEVEL IN LEAF TISSUES OF WOODY PLANTS DURING ABSCISSION

JOSEPH RIOV, ODED SAGEE and RAPHAEL GOREN

Department of Horticulture, Faculty of Agriculture, The Hebrew
University of Jerusalem, Rehovot 76100, Israel

The ability of exogenously supplied ethylene to promote abscission
of intact leaves varies greatly among different species (1). This is
particularly true for woody plant species where some abscise one or
two days after the beginning of treatment with ethylene while others
respond to the hormone after much longer periods. Although it is well
accepted that auxin-ethylene interactions control abscission processes
(1,3), there are no conclusive data to explain why various species
differ greatly in their response to ethylene. Beyer & Morgan (5)
proposed a model for the role of ethylene in the regulation of
abscission of intact leaves. The proposed regulatory system first
involves a modification of the hormonal balance in the abscission zone
achieved as ethylene reduces indole-3-acetic acid (IAA) transport
capacity of the petioles. After the reduction of auxin levels, ethylene
exerts direct action in the abscission zone such as stimulating synthesis
of cell-wall degrading enzymes (1,3) and secretion of these enzymes into
the cell-wall (2). Also, ethylene has been demonstrated to reduce
endogenous IAA level (4,6,9) and there is some evidence that ethylene
may reduce auxin level in the abscission zone either by stimulating
destruction (6,7,10,11,15), or by inhibiting synthesis (6). Beyer (4)
demonstrated that the leaf blade is the initial target tissue of
exogenously supplied ethylene, where some essential function of the

Y. Fuchs and E. Chalutz (eds.) Ethylene: Biochemical, Physiological and Applied Aspects.
ISBN 90-247-2984-X. Printed in The Netherlands

hormone must first be performed before abscission can occur. There are data which indicate that this essential function of ethylene is to reduce the amount of auxin transported out of the blade, possibly by reducing auxin levels and inhibiting auxin transport in the veinal tissues (4).

In the recent years we have conducted a research aimed to study whether the differences in the abscission response of various woody plants to ethylene might be related to the effects of the hormone on IAA transport and level of IAA in leaf tissues. We compared the behavior of a sensitive (citrus) and insensitive (eucalyptus) species to abscission induction by ethylene. In this paper we summerize the data accumulated in our laboratory on this subject.

EFFECT OF ETHYLENE ON IAA TRANSPORT

Inhibition of auxin transport seems to be one of the main effects of ethylene at an early stage of the abscission process (5). We examined the effect of ethylene on IAA transport in midrib tissue of citrus and eucalyptus as related to their abscission response (15). IAA transport was measured by the classical donor-receiver agar cylinder technique. The ability of ethylene to promote leaf abscission differed greatly among these species (Figs. 1 and 2). Whereas citrus leaves reached 100% abscission after 48 hours of ethylene treatment (Fig. 1), abscission of eucalyptus leaves started between 96 and 120 hours after the beginning of treatment and reached about 70% after 168 hours (Fig. 2).
In both species, abscission of ethylene-treated leaves showed a kinetic trend similar to that of the inhibition of auxin transport (Figs. 1 and 2). As might be expected the inhibition of auxin transport preceded abscission. In citrus leaves, ethylene induced a rapid and marked reduction in IAA transport, reaching about 80% inhibition after 24 hours

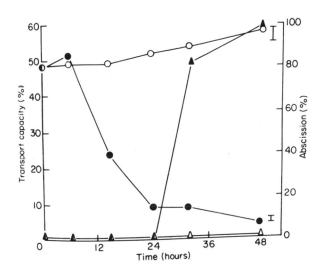

FIGURE 1. Effect of ethylene on abscission and IAA transport in midrib sections of citrus leaves. Air: (△) abscission; (○) transport capacity. Ethylene: (▲) abscission; (●) transport capacity.

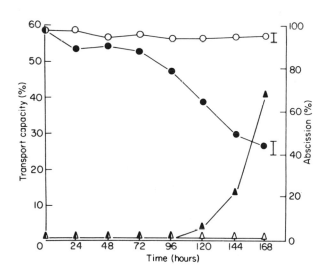

FIGURE 2. Effect of ethylene on abscission and IAA transport in midrib sections of eucalyptus leaves. Key as in Figure 1.

of exposure (Fig. 1). The reduction of IAA transport in eucalyptus mid-rib sections was much slower and less pronounced (Fig. 2). A significant inhibition of IAA transport was observed only after a 96-hour ethylene pretreatment, and reached 50% after 144 hours. The data suggest that as long as the basipetal flow of IAA from the leaf blade continues unaltered, ethylene cannot induce abscission. The ability of ethylene to inhibit auxin transport may therefore be an important factor in determining the sensitivity to ethylene.

EFFECT OF ETHYLENE ON METABOLISM OF (^{14}C) IAA

An additional factor which may determine the sensitivity to ethylene is the ability of the hormone to reduce the level of IAA in the leaf blade which is the site of IAA synthesis (4). Several papers demonstrate that ethylene reduces auxin level in plant tissues (4,6,9). This reduction may result from increased metabolism (6,7,10,11,15) and/or inhibition of synthesis (6). We compared the effect of ethylene on metabolism of exogenously supplied ($2-^{14}$C)IAA in leaf tissues of citrus and eucalyptus (14).

Leaf discs or midrib sections were cut from leaves preincubated in either air or ethylene. After incubation in labeled IAA, the tissue was subjected to differential extraction of the radioactivity with 80% ethanol, water and 1N NaOH. Data are presented only for 1 hour of incubation in labeled IAA since it is believed that the initial rate of conjugation represents more closely the in vivo situation.

Twenty four hours of ethylene pretreatment significantly increased IAA conjugation in citrus leaves, but had almost no effect on IAA metabolism in eucalyptus tissues (Table 1). Of particular interest is the 4-fold increase in the radioactivity in the NaOH fraction of ethylene pretreated citrus leaves.

The water fraction contained a small amount of radioactivity which showed the same pattern of changes as the ethanol. The data indicate that in citrus leaf tissues ethylene may, besides inhibiting IAA transport, reduce the level of endogenous IAA by increasing its metabolism, whereas IAA metabolism in eucalyptus leaves is almost unaffected by ethylene.

Table 1. Initial rate of (^{14}C)IAA metabolism in leaf tissues of citrus and eucalyptus.

Species	Tissue	Air		Ethylene	
		Ethanol fraction	NaOH fraction	Ethanol fraction	NaOH fraction
		%*		%	
Citrus	Leaf discs	14	1.4	24	5.9
	Midrib sections	28	1.9	43	9.4
Eucalyptus	Leaf discs	17	1.5	18	1.9
	Midrib sections	26	0.9	27	1.6

* Bound IAA as percentage of total uptake.

NATURE OF IAA METABOLITES

Most studies of the effects of ethylene on IAA metabolism have dealt with metabolites soluble in 80% ethanol (6,10,15). We observed that a relatively high amount of radioactivity remained in the plant residue of ethylene-pretreated tissues after extraction with 80% ethanol as compared to air controls. The extraction procedure utilized enabled us to extract most of the radioactivity from the tissue and also afforded some separation between the IAA metabolites. Thin-layer chromatography of the ethanol and water extracts of both citrus and eucalyptus tissues showed that these fractions contained low molecular weight conjugates. In citrus, ethylene pretreatment not only increased the rate of conjugation but also altered the nature of the conjugates. The major conjugate in air control was identified by gas chromatography - mass spectrometry as IAA-aspartate. Ethylene pretreatment reduced the formation of IAA-aspartate and very significantly increased the formation of a neutral conjugate, probably a glycosyl ester. Only trace amounts of the neutral conjugate could be detected in extracts of air controls. No qualitative changes in low molecular weight conjugates were observed in eucalyptus tissues following ethylene treatment.

Chromatography of the NaOH extracts of citrus tissues on a Sephadex G-25 column (Fig. 3) yielded material characterized by one major peak which coincided with the protein peak and several minor ones. Radioactivity in all peaks and particularly in the major one was much higher in the ethylene pretreatment compared to the control. Inasmuch as the major peak was excluded from the Sephadex G-25 column, the molecular weight of the IAA conjugate(s) in this fraction is higher than 5,000.

Differential centrifugation of the ethanol - and water-insoluble material revealed that most radioactivity was found in the 5,000 g pellet (Table 2), indicating that it was associated with the cell-wall. Further studies of the nature of these bound forms of IAA was performed by extracting the ethanol- and water- insoluble material with different solvents according to Scheel and Sandermann (16) with some modifications. The procedure utilized successive extractions with sodium dodecylsulphate (SDS), dimethylformamide (DMF), dioxane/water, and 1N NaOH. SDS has been shown to extract mainly proteins whereas DMF and dioxane are commonly used for extraction of lignin. Significant amount of radioactivity were detected in all extracts, suggesting that IAA is probably

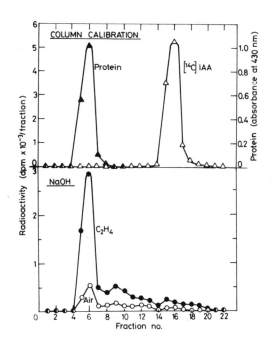

FIGURE 3. Sephadex G-25 chromatography of NaOH-extractable IAA conjugates of citrus leaf discs from control and ethylene-pretreated leaves.

bound to different cell-wall components. Macromolecular conjugates of IAA have been described in various plants (cf. 14). These conjugates include proteins, polysaccharides, and gluco-proteins. The exact nature of the macromolecular weight conjugates of citrus tissues is not yet clear. Nevertheless, this type of binding may serve as an inactivating system of IAA.

Table 2.   Distribution of ethanol – and water-insoluble radioactivity among subcellular fractions.

	A i r		Ethylene	
	dpm	%	dpm	%
Ethanol – and water– insoluble	4,190	100	38,994	100
5,000 g	3,270	78	33,873	87
10,000 g	420	10	2,775	7
100,000 g	500	12	2,346	6

IAA LEVEL IN THE ABSCISSION ZONE

The data presented above suggest that ethylene may reduce the level of endogenous IAA in the abscission zone either by inhibiting transport from the site of synthesis and/or increasing conjugation. To determine changes in IAA level in the abscission zone during the abscission process, we developed a specific enzyme immunoassay for determination of IAA. Antiserum for IAA was obtained from rabbits immunized with IAA bound to bovine serum albumin through the indole nitrogen by the Mannich reaction (13). The assay proved to be highly specific for IAA. Other indoles structurally related to IAA were cross reactive only at a very high concentration. Before analysis IAA has to be purified to remove interfering compounds. Purification was achieved by insoluble polyvinylpyrrolidone, DEAE-Sephadex, and C-18 column chromatography.

Data for changes in IAA level in the abscission zone of citrus leaf explants although preliminary in nature, showed that the level of IAA increased following excision. The increase was more rapid in ethylene-treated explants. Later on IAA level decreased markedly in the latter explants, whereas the level in air-treated explants remained high up to

24 hours after excision. A similar pattern of changes of auxin-like activity in the abscission zone of citrus leaf explants was reported (7).

The changes in IAA level in the abscission zone of citrus leaf explants seems to fit with the two-phase sequence in abscission (3). The first phase is in fact a growth process which involves mainly cell enlargement (12). Such growth process also occurs in the abscission zone of citrus leaf explants prior to abscission (8). Although cell enlargement during abscission has been related to ethylene (17), it may be that at least in some species the initial accumulation of IAA is responsible for this phenomenon. Ethylene stimulates abscission by reducing IAA level in the second stage, thus exposing the abscission zone to the direct action of ethylene. The data presented for citrus suggest that this reduction may result from both the inhibition of IAA transport and stimulating its metabolism (Fig. 1, Table 1). We have yet to determine (a) if similar changes in IAA level also occur in the abscission zone of intact citrus leaves and (b) what are the changes of IAA level in the abscission zone of eucalyptus leaves.

REFERENCES

1. Abeles FB. 1973. Ethylene in Plant Biology. Academic Press, New York.
2. Abeles FB, Leather GR. 1971. Planta 97:87-91.
3. Addicott FT. 1982. Abscission. University of California Press, Berkeley and Los Angeles.
4. Beyer EM Jr. 1975. Plant Physiol. 55:322-327.
5. Beyer EM Jr. Morgan PW. 1971. Plant Physiol. 48:208-212.
6. Ernest LC, Valdovinus JG. 1971. Plant Physiol. 48:402-406.
7. Gaspar T, Goren R, Huberman M, Dubucq M. 1978. Plant, Cell Environ. 1:225-230.
8. Jaffe MJ, Goren R. 1979. Bot. Gaz. 140:378-383.
9. Lieberman M, Knegt E. 1977. Plant Physiol. 60:475-477.
10. Minato T, Okazawa Y. 1978. J. Fac. Agric. Hokkaido Univ. 58:535-547.
11. Morgan PW, Beyer EM Jr, Gausman HW. 1968. In: Biochemistry and Physiology of Plant Growth Substances, (eds. F. Wightman, G. Setterfield), pp. 1255-1273. Runge Press, Ottawa.
12. Osborne DJ. 1973. In: Shedding of Plant Parts, (ed. T.T. Kozlowski), pp, 125-147. Academic Press, New York and London.

13. Pengelly W, Meins F Jr. 1977. Planta 136:173–180.
14. Riov J, Dror N, Goren R. 1982. Plant Physiol. 70:1265–1270.
15. Riov J, Goren R. 1979. Plant, Cell Environ. 2:83–89.
16. Scheel D, Sandermann H Jr. 1981. Planta 152:253–258.
17. Wright M, Osborne DJ. 1974. Planta 120:163–170.

EFFECTS OF THE DEFOLIANT THIDIAUZURON ON LEAF ABSCISSION AND
ETHYLENE EVOLUTION FROM COTTON SEEDLINGS

J.C. SUTTLE
USDA-ARS, Metabolism and Radiation Research Lab., Fargo, ND

INTRODUCTION
     Chemical defoliants have become an integral aspect of
modern cotton production and are currently used on an estimated
75% of total cotton acreage in the United States. Thidiazuron
is currently registered for use in the U.S. as a cotton
defoliant. The abscission promoting properties of this compound
were first described in 1976 (1), but its mode-of-action
remains unknown. Preliminary experiments in this laboratory
have shown that treatment of cotton seedlings with thidiazuron
results in an elevated rate of ethylene evolution (2). Ethylene
is currently regarded as an endogenous regulator of abscission
in many higher plants (3). The data presented herein describe
a portion of an on-going research program dealing with the
physiological effects of chemical defoliants.

MATERIALS AND METHODS
     Cotton seedlings (Gossypium hirsutum L. cv. Stoneville
519) were used in all studies. Seedlings were used when the
fourth true leaf was initiating expansion. Abscission-zone
explants were prepared from the cotyledonary nodes of these
seedlings. The isolated explants were surface-sterilized and
were manipulated under aseptic conditions. Thidiazuron
(N-phenyl-N'-1,2,3-thiadiazol-5-ylurea) was provided by
Nor-Am Agricultural Products, Inc. Ethylene was determined
by gas chromatography.

RESULTS
     Treatment of cotton seedlings with concentrations of thidi-
azuron (TDZ) in excess of 1×10^{-5} mol dm^{-3} resulted in greater

Y. Fuchs and E. Chalutz (eds.) Ethylene: Biochemical, Physiological and Applied Aspects.
ISBN 90-247-2984-X. Printed in The Netherlands
©1984, Martinus Nijhoff/Dr W. Junk Publishers, The Hague.

than 80% abscission of young, expanding leaves and only 30%
abscission of mature leaves. Abscission of younger leaves
could be detected 2 days post-treatment and was complete within
5 days. Elevated rates of ethylene evolution could be detected
from both leaf blades and abscission-zone explants isolated
from TDZ-treated seedlings. This stimulation was evident
within 24 hr of treatment and was greatest in mature leaf
tissues ($>$ 8.9 nmol/gr.f.wt./hr). When applied directly to
isolated abscission-zone explants, TDZ stimulated the rate of
ethylene evolution while retarding the rate of abscission.
Silver thiosulfate completely inhibited TDZ-stimulated
abscission in intact seedlings.

DISCUSSION

The fact that increased ethylene evolution preceeded the
onset of abscission in TDZ-treated tissues coupled with the
ability of silver thiosulfate to inhibit this stimulation
suggests that ethylene mediates at least a portion of the
mode-of-action of this defoliant. A more definitive statement
concerning ethylene's possible role clearly awaits further
evidence. The inability of TDZ to accelerate abscission when
applied directly to abscission-zone explants indicates that
the leaf tissue is the principle physiological target of this
defoliant.

REFERENCES
1. Arndt F, Rusch R, Stilfried HV. 1976. Plant Physiol
   57: S-99.
2. Suttle JC. 1983. Plant Physiol 72: S-121.
3. Jackson MB, Osborne DJ. 1970. Nature 225: 1019.

ETHYLENE AND AUXIN TRANSPORT, AND METABOLISM IN PEACH FRUIT ABSCISSION

A.RAMINA, A.MASIA, AND G.VIZZOTTO
Institute of Pomology, University of Padova, 35100 Padova, Italy

The purpose of the present study was to evaluate in peach fruit a possible interaction of ethylene and embryoctomy with auxin transport and metabolism through the abscission zone located between fruit and receptacle (AZ3).

Material and methods

Plant material and treatments. Research was carried out on fruits of 12 year old 'Andross' peach trees grown at Ferrara (Italy). At the end of stage I of fruit development the following treatments were applied: CEPA to the fruit (CEPA$_f$), CEPA to the pedicel (CEPA$_p$), 2,3,5-triiodobenzoic acid (TIBA) to the pedicel (TIBA$_p$), TIBA to the pedicel and CEPA to the fruit (TIBA$_p$+CEPA$_f$), embryoctomy. One group was kept intact as control. The CEPA (100ppm), and TIBA (1%) solutions were smeared on the fruit and/ /or pedicel surface using a brush. Embryoctomy was done by cutting the fruit transversally and removing the seed.

Auxin transport and metabolism. Immediately after imposition of treat ments, and 8, 24, 48 hr later, explants including AZ3 cells were excised. Auxin transport and metabolism was studied by using the agar block technique according to Weinbaum et al (3). Donor blocks containing 1.5% agar and 20 µM 1-^{14}C-IAA (specific radioactivity 60mCi/mM) were placed in contact with the explants and incubated for 6 hr in the dark at 25°C. Explants were subjected to differential extraction according to Davies (1). Analysis of EtOH soluble metabolites was carried out on TLC using chlorophorm-ethylacetate-formic acid (5:4:1,v/v) as solvent.

Results and discussion

Abscission. Kinetics of abscission were differently affected by treatments: embryoctomy induced a quick activation of AZ3 cells, and all fruits abscised in 1 week; CEPA$_f$,CEPA$_f$+TIBA$_p$, and TIBA$_p$ were equally effective in inducing, at the end of june drop, % of abscission statisti-

Y. Fuchs and E. Chalutz (eds.) Ethylene: Biochemical, Physiological and Applied Aspects.
ISBN 90-247-2984-X. Printed in The Netherlands

cally equal although kinetics were different. CEPA$_p$ failed in promoting abscission and the treated fruits displayed kinetics and total abscission equal to the control.

1-^{14}C-IAA transport. The greatest reduction of 1-^{14}C-IAA transport was induced by embryoctomy. TIBA$_p$, CEPA$_f$, and TIBA$_p$+CEPA$_f$ were also quite effective in reducing transport, while CEPA$_p$ reduced translocation only at time 0.

Solvent distribution. In explants excised from control and CEPA, above 80% of the radioactivity was recovered in the EtOH fraction; in other treatments the radioactivity present in the same fraction progressively declined with time, reaching the lowest value (53.4%) in the case of CEPA$_f$+TIBA$_p$, 48 hr after imposition of treatment. Radioactivity recovered in H$_2$0 ranged between 3.1 to 8.2%. The highest values were observed in explants excised from control and CEPA$_p$ 48 hr after imposition of treat-ment. Radioactivity in the NaOH extract and residue of both the control and the CEPA$_p$ remained low through the experimental period, while in explants from other treatments increased with time.

EtOH-extractable metabolites. TLC of EtOH extracts revealed the presence of three main zones of radioactivity in all treatments, identified by co-chromatography with standards as IAA, IAglu and IAAsp. Explants excised from control and CEPA$_p$ presented the lowest amount of free IAA and the highest levels of IAAsp. Among other treatments, TIBA$_p$ and TIBA$_p$+CEPA$_f$ were the most effective in blocking IAA conjugation to IAAsp.

REFERENCES

1. Davies PJ. 1976. Bound auxin formation in growing stem. Plant Physiol 57:197-202.
2. Ramina A, Giulivo C, Rascio N, and Casadoro G. 1982. Natural and 2-chloroethylphosphonic acid (CEPA) -induced fruit abscission in Prunus persica L. Batsch at electron microscopy level. XXIst Intern. Hort. Congress, Hamburg abs 1067.
3. Weinbaum SA, Giulivo C, Ramina A. 1977. Chemical thinning: ethylene and pre-treatment fruit size influence enlargement, auxin transport, and apparent sink strenght of french prune and 'Andross' peach. J. Am. Soc. Hort. Sci. 102:781-785.

ETHYLENE AND THE CONTROL OF TOMATO FRUIT RIPENING

GRAEME E. HOBSON, JANE E. HARMAN and ROYSTON NICHOLS

Glasshouse Crops Research Institute, Worthing Road, Littlehampton,
W. Sussex, U.K.

The sequence of events collectively known as ripening is the single
most dramatic event in the life of climacteric fruit. It is quite clear
that in the tomato, ribonucleic acid (1) and protein synthesis (see 2)
are involved in the process, and a great deal of study has gone into
an elucidation of the sequence of changes, especially in terms of precise
alterations in enzymic components of various fruits (3-7). While specific
proteins play a part in promoting ripening, it is possible that increased
turnover, activation and transfer of proteins across membranes can best
explain the process, possibly with ethylene being concerned with the
initiation and co-ordination of many of the separate events. Some of
the more obvious and easily followed changes are illustrated in Figure 1.

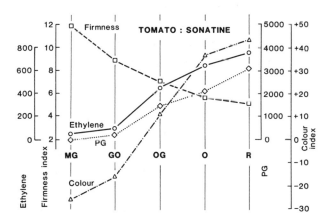

FIGURE 1. The rate of change of four parameters as mature green tomato
fruit ripen. The rise in ethylene precedes that for the enzyme polygalacturonase (PG) by 24-48 h (8, 9).

How can we use the techniques at our disposal to understand more
about the role of ethylene in the ripening of climacteric fruit such
as the tomato, and how can the rate of ethylene synthesis be manipulated

Y. Fuchs and E. Chalutz (eds.) Ethylene: Biochemical, Physiological and Applied Aspects.
ISBN 90-247-2984-X. Printed in The Netherlands
© 1984, Martinus Nijhoff/Dr W. Junk Publishers, The Hague.

to give some control over the postharvest life of the fruit?

One of the first events that heralds the onset of ripening is an increase in ethylene production and respiration from a low steady level. Polygalacturonase (PG), an enzyme that appears to be intimately concerned with texture changes (8) and whose natural substrate is pectic acid, is not active in green tomatoes but is synthesized soon after rises in ethylene are detected (9, 10). The enzyme exists in several multimolecular forms, with component 1 appearing first and having about twice the molecular weight of two more components, known as 2A and 2B (10). There is evidence that these isoenzymes are all variants of the same polypeptide (9). Pectinesterase activity appears not to be limiting throughout ripening, but its action is a pre-requisite for efficient degradation by PG. While induction of PG by ethylene has not been demonstrated, the view can be held that until the autocatalytic phase of ethylene production is seen, little or no PG synthesis takes place. Also an increasing number of cases can be quoted where the ratio of maximum ethylene production to the basal rate is positively correlated with PG activity at full development (see 7).

There is convincing evidence from low-pressure storage experiments with tomatoes that below a critical level of ethylene and oxygen (see 7), where ethylene-binding sites are altered (11, vide infra), or where tissue from parts of fruit is not appropriately conditioned to respond to ethylene through chilling injury or physiological disorder (12), ripening is either inhibited or irreversibly prevented. A number of single gene mutations

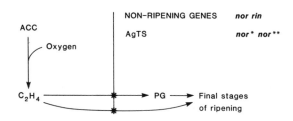

CONTROL OF RIPENING

FIGURE 2. A representation of the inhibition of polygalacturonase (PG) and other enzymes involved in ripening by 'non-ripening' genes or silver thiosulphate (AgTS). The signs x and xx indicate mutants such as Spanish Winter, Longkeeper and Alcobaca which may be modifications at the non-ripening locus.

of tomato plants have the effect of preventing ripening from taking place at the usual time for full development (13). Similarly, the introduction of about 1 µmole of silver thiosulphate (AgTS) through the transpiration stream into intact tomatoes interferes with the ability of parts of the walls of the fruit to ripen (14). The situation is illustrated in Figure 2.

Neither the genetic nor the chemical inhibition of ripening can be overcome by exposure to physiological levels of ethylene (see 13). However, attached fruit containing the 'ripening inhibitor' (rin) gene can be partly ripened by exposure to excess ethylene (15), while detached fruit of this line require both ethylene and oxygen (16). Following the work of Mizrahi and his group (17), we have confirmed that salt stress can overcome to some extent the block to ripening by some of the mutant genes, especially in combination with application of an ethylene-producing chemical. Typical effects on PG activity, colour, ethylene evolution and firmness of the fruit are shown in Table 1. The stimulation of ripening

Table 1. Response by tomato mutants to salts (500 ml solution per day) and 'ethrel' (0.1% painted on mature fruit every third day)

Genotype and treatment	Units of PG[a]	Tomato colour index[b]	Ethylene $(nl\ g^{-1}h^{-1})$	Firmness index[c]
Alcobaca (non-isogenic with cv. Ailsa Craig)				
Control	48.4	-15.51	47	9.09
0.5% NaCl	60.5	15.51	97	6.29
0.5% NaCl + 'ethrel'	51.2	16.57	288	5.88
Spanish Winter (non-isogenic with cv. Ailsa Craig)				
Control	114.4	7.61	337	4.65
0.5% NaCl	214.8	43.69	511	3.77
0.5% NaCl + 'ethrel'	184.7	41.19	404	3.57
Non-ripening nor nor (isogenic with cv. Ailsa Craig)				
Control	7.4	-14.91	—	6.75
1% NaCl	52.9	15.01	—	4.42
1% Na_2SO_4	8.2	18.10	—	5.03

[a]See (18); mg galacturonic acid released /h/100 g fresh wt.

[b]See (19)

[c]Reciprocal of the compression in cm under a load of 1 kg for 5 sec

is not confined to sodium chloride since several other salts that were taken up by the fruit were also effective. In general, salts affected the colour and firmness much more than either PG or ethylene levels. In Alcobaca and nor nor lines, only PG1 was found; Spanish Winter showed a small amount of PG2A and 2B as well. The very long shelf-life and general resistance to deterioration shown not only by salt-treated non-ripening mutants but also by hybrids containing one or more non-ripening alleles (7, 13) appear to be a reflection of PG activity that is much lower than normal. Ethylene levels are also attenuated, and in one

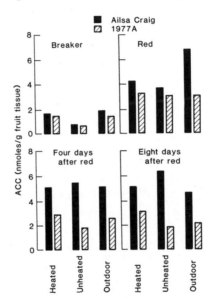

FIGURE 3. 1-aminocyclopropane-1-carboxylic acid (ACC) levels in cv. Ailsa Craig (++) and in an F_1 hybrid containing the rin allele (rin +) (1977A)

instance (Figure 3) has been shown to be matched by reduced levels of ACC (Mordy Atta Aly, unpublished). In summary, the ability of homozygous non-ripening lines to enter the autocatalytic phase of ethylene production is clearly inhibited. In some cases, the block to further development may be overcome but whatever the stimulus used, PG activity and ethylene levels are much below normal.

Of the well-known ethylene antagonists that we have infiltrated into mature green tomato fruit detached from the plant immediately before treatment, 2,5-norbornadiene (20), α-aminooxyacetic acid (21), α-amino-

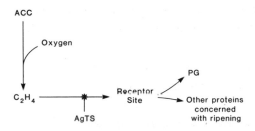

FIGURE 4. In the inhibition of ripening by silver thiosulphate, the assumption is that the complex attaches itself to a site in the tissue that would have responded to ethylene

isobutyric acid (22) and AgTS, the silver salt was by far the most effect-ive in preventing ripening. In further experiments using unpicked green tomatoes, infiltration of 1 μmole of AgTS into the vascular system in part of the peduncle leading to a fruit truss caused a proportion of the outer walls to fail to change colour. The composition and pectic enzyme activities in both the green and red walls of treated fruit (unpublished data) can be compared with normal tissue having similar colour. In essence, the composition of all wall tissue following silver infiltration was abnormal, and was generally similar to that in fruit showing a

Table 2. Characteristics of outer locule wall tissue from fruit infiltrated with silver thiosulphate (cv. Sonatine)

Source of tissue	Units of PG[a]	Tomato Colour Index[b]	Ethylene ($nl\ g^{-1}\ h^{-1}$)
Control fruit			
Mature green	0	−24.43	0.4
Red	6531	46.18	8.6
Silver-treated fruit			
Green	226	−17.48	4.4
Red	4753	51.61	11.6

[a]See (18); mg galacturonic acid released/h/100 g fresh wt.

[b]See (19)

ripening disorder known as 'blotch' (23). As indicated in Table 2, the green tissue from infiltrated fruit failed to change colour much, and although apparently producing adequate amounts of ethylene, less than 5% of the PG activity of red tissue from the same fruit was shown. In order to prevent ripening, the silver treatment must be given prior to the autocatalytic phase of ethylene production.

The green tissue from infiltrated fruit contained particles that were electron-dense when sections were examined by electron microscopy. The deposits were found in phloem cell walls and were particularly associated with the middle lamella. In addition, the particles were observed in the intercellular spaces of parenchyma, bordering the lumen, as is shown in

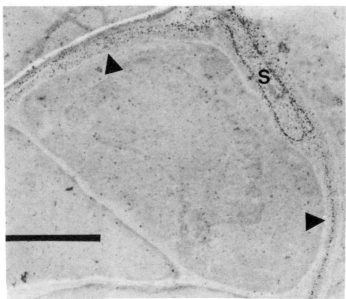

FIGURE 5. Particles (arrowed) containing concentrations of silver and sulphur deposited throughout the cell wall of phloem, but more concentrated in the middle lamellae and intercellular space (S), in tomato tissue infiltrated with 20 mM silver thiosulphate. Lower concentrations of the silver complex resulted in particles being found more exclusively in the middle lamellae. Bar mark = 2 μm; unstained

Figure 5. X-ray microanalysis has confirmed that the particles contained concentrations of silver and, to a lesser extent, sulphur (unpublished data).

Since AgTS infiltration resulted in a failure by the green tissue to synthesize PG in normal quantities, we have attempted to show that the tissue was still capable of making protein. No evidence of silver-containing deposits were found in the cytosol. Preliminary work indicates that an inducible enzyme such as nitrate reductase was synthesized in green tissue from silver-treated fruit. Nevertheless, conclusive proof must await further tests to show that amino acids can be incorporated into protein in this tissue.

CONCLUSION

Silver is a particularly potent ion in preventing the ripening of tomato tissue. The metal has strong anti-ethylene properties (see 24), and if it binds to those sites that are concerned with increasing sensitivity towards, or with the increased production of the hormone, this could prevent normal ripening through inhibiting the synthesis of PG and other proteins concerned with completion of ripening. A direct effect of silver on the protein synthetic mechanism is not thought likely.

In normal ripening, rising ethylene levels are closely followed by selective protein synthesis, of which the enzyme PG is an important example. Whether PG is induced by ethylene is at present an open question. However, it is clear that the ripening mechanism is extremely sensitive to absolute levels of ethylene and to an activation mechanism for recognition of the hormone. Further investigations into the biochemical causes for natural (e.g., 'blotch', non-ripening mutants) or deliberate interference with the ripening sequence may help to pinpoint the controlling steps in the sequence of changes. A more complete knowledge of the processes involved should diminish spoilage of tomato fruit, and contribute towards economic benefit for both producers and consumers.

ACKNOWLEDGEMENTS

We would like to acknowledge the contributions made by Messrs P. Atkey, J. Pegler, Mordy Atta Aly, Dr A. Sharaf, Mrs C. Richardson and Mrs C. Frost to the results reported here. It is also a pleasure to record our thanks to Dr A. Brown, Jeol (U.K.) Ltd. for expert help and advice with the X-ray microanalysis.

REFERENCES

1. Grierson D. 1983. Control of ribonucleic acid and enzyme synthesis during fruit ripening. In NATO Advanced Study Institute on Post-harvest Physiology and Crop Protection, Sounion, Greece, 1981 (M. Lieberman, ed.). New York, Plenum Press.
2. Tucker GA, Grierson D. 1982. Synthesis of polygalacturonase during tomato fruit ripening. Planta 155, 64-67.

3. Sacher JA. 1973. Senescence and postharvest physiology. Annu. Rev. Plant Physiol. 24, 197-224.
4. Hobson GE. 1974. Electrophoretic investigation of enzymes from developing Lycopersicon esculentum fruit. Phytochem. 13, 1383-1390.
5. Tucker GA, Robertson NG, Grierson D. 1980. Changes in polygalacturonase isoenzymes during the ripening of normal and mutant tomato fruit. Europ. J. Biochem. 112, 119-124.
6. Tucker GA, Robertson NG, Grierson D. 1982. Purification and changes in activities of tomato pectinesterase isoenzymes. J. Sci. Food Agric. 33, 396-400.
7. Davies JN, Hobson GE. 1981. The constituents of tomato fruit - the influence of environment, nutrition and genotype. CRC Crit. Rev. Food Sci. & Nutr. 15, 205-280.
8. Hobson GE. 1981. Enzymes and texture changes during fruit ripening. In Recent Advances in the Biochemistry of Fruit and Vegetables (J. Friend & M.J.C. Rhodes, eds.). Phytochemical Society of Europe, pp. 121-130. London, Academic Press.
9. Grierson D, Tucker GA. 1983. Timing of ethylene and polygalacturonase synthesis in relation to the control of tomato fruit ripening. Planta 157, 174-179.
10. Ali ZM, Brady CJ. 1982. Purification and characterisation of the poly-galacturonases of tomato fruits. Aust. J. Plant Physiol. 9, 155-169.
11. Sisler EC. 1982. Ethylene binding in normal, rin and nor mutant tomatoes. J. Plant Growth Regul. 1, 219-226.
12. Hobson GE, Davies JN, Winsor GW. 1978. Ripening disorders of tomato fruit. Growers Bulletin No. 4, Glasshouse Crops Research Institute, Littlehampton, West Sussex.
13. Tigchelaar EC, McGlasson WB, Buescher RW. 1978. Genetic regulation of tomato fruit ripening. HortScience 13, 508-513.
14. Hobson GE, Nichols R, Aly MA. 1983. Inhibition of tomato fruit ripening by silver, Plant Physiol. 72, S-168.
15. Mizrahi Y, Dostal HC, Cherry JH. 1975. Ethylene-induced ripening in attached rin fruits, a non-ripening mutant of tomato. HortScience 10, 414-415.
16. Frenkel C, Garrison SA. 1976. Initiation of lycopene synthesis in the tomato mutant rin as influenced by oxygen or ethylene inter-actions. HortScience 11, 20-21.
17. Mizrahi Y, Zohar R, Malis-Arad S. 1982. Effect of sodium chloride on fruit ripening of the non-ripening tomato mutants nor and rin. Plant Physiol. 69, 497-501.
18. Hobson GE. 1980. Effect of the introduction of non-ripening mutant genes on the composition and enzyme content of tomato fruit. J. Sci. Food Agric. 31, 578-584.
19. Hobson GE, Adams P, Dixon TJ. 1983. Assessing the colour of tomato fruit during ripening. J. Sci. Food Agric. 34, 286-292.
20. Sisler EC, Yang SF. 1983. Effect of butenes and cyclic olefines on etiolated pea plants in relation to the ethylene response. Plant Physiol. 72, S-40.
21. Amrhein N, Wenker D. 1979. Novel inhibitors of ethylene production in higher plants. Plant & Cell Physiol. 20, 1635-1642.
22. Satoh S, Esashi Y. 1980. α-Aminoisobutyric acid: a probable competitive inhibitor of conversion of 1-aminocyclopropane-1-carboxylic acid to ethylene. Plant & Cell Physiol. 21, 939-949.
23. Winsor GW, Massey DM. 1959. The composition of tomato fruit. II. Sap expressed from fruit showing colourless areas in the walls. J. Sci.

Food Agric. <u>10</u>, 304-307.

24. Veen H. 1983. Silver thiosulphate: an experimental tool in plant
    science. Scientia Hort. <u>20</u>, 211-224.

EXPERIMENTS TO PREVENT ETHYLENE BIOSYNTHESIS AND/OR ACTION
AND EFFECTS OF EXOGENOUS ETHYLENE ON RIPENING AND STORAGE OF
APPLE FRUITS 1)

F. BANGERTH, G. BUFLER AND H. HALDER-DOLL 2)

Universität Hohenheim, Institute für Obst-Gemüse und Weinbau,
7000 Stuttgart 70, Germany

Ethylene is considered to be one of the dominating hormones
in controlling ripening in climacteric fruits like apples. However, it is difficult to elucidate the regulatory function of
ethylene in the various biochemical processes of fruit ripening in the presence of endogenous ethylene. In order to prevent apples from entering the stage of autocatalytic ethylene
production, two methods were used:
- storage under low or hypobaric pressure (LPS) as described
  by Burg and Burg (1966).
- treatment of apple fruits on the tree with the ethylene
  biosynthesis inhibitor AVG (aminoethoxyvinylglycine), introduced by Lieberman et al. (1975).
These methods allow the effect of exogenous ethylene on various ripening parameters of apple fruits to be studied. Some
relevant experiments done in our laboratory will be presented
and discussed below.

MATERIAL AND METHODS
Apple fruits of the cv. "Golden Delicious" from the Experiment Station of the University of Hohenheim (FRG) were treated with AVG 4 to 6 weeks before the assumed harvest date.
The treatment was repeated 2 to 3 weeks later. AVG- or untreated control (Co.) fruits were harvested in a preclimacteric
stage as judget by endogenous C_2H_4 concentration. They were

---

1) We dedicate this publication to the memory of Dr. M. Lieberman
2) Names in alphabetical order

Y. Fuchs and E. Chalutz (eds.) Ethylene: Biochemical, Physiological and Applied Aspects.
ISBN 90-247-2984-X. Printed in The Netherlands
©1984, Martinus Nijhoff/Dr W. Junk Publishers, The Hague.

either used immediately for experiments at 20 or 25°C or stored at 3°C under normal or hypobaric pressure.

RESULTS AND DISCUSSION

Effect of C_2H_4 on the regulation of its own biosynthesis
Ethylene is synthesized from methionine via S-adenosylmethionine and l-aminocyclopropane-l-carboxylic acid (ACC) (Adams
and Yang 1979). ACC synthase is considered to be a rate-limiting enzyme in this pathway (Yang 1980). No ACC synthase activity was detectable in apples stored continuously for 140
days under low pressure (Fig. 1) and ACC content and ethylene

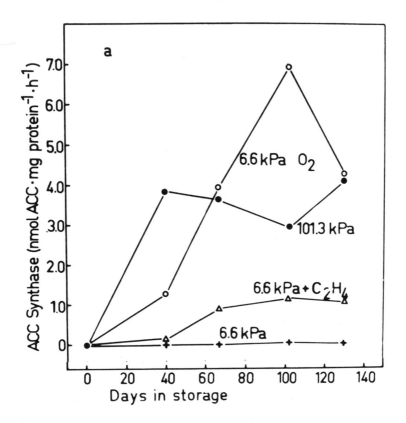

Fig. 1: Effects of partial pressures of oxygen (O_2) and ethy-
lene (C_2H_4) on ACC-synthase activity of apples stored under
LPS (6.6 kPa, 4°C) or at normal (101.3 kPa) pressure (from
Bufler and Bangerth 1983).

production did not increase (Bufler and Bangerth 1983). Under hypobaric storage conditions the partial pressure of oxygen is reduced and endogenously produced ethylene is continuously removed. If apples, stored under these conditions, were continuously treated with ethylene, ACC synthase activity was induced (Fig. 1), leading to an increase in ACC content. Treatment with propylene also induced ACC synthase activity (Bufler and Bangerth 1983). If the partial pressure of oxygen in the LPS storage container was increased by ventilation with oxygen, ACC synthase activity developed, although not as fast as in fruits stored at normal pressure (Fig. 1).

The described experiments suggest induction of ACC synthase activity by ethylene. In order to investigate whether ACC synthase activity is dependent on ethylene concentration, AVG-treated fruits were ripened at 25°C with 50 $\mu l \cdot l^{-1}$ exogenous ethylene. After stopping ethylene treatment for four days, the ethylene-induced ACC synthase activity declined to non-detectable amounts (Bufler 1984). Apples were then treated with different concentrations of ethylene. At 0.07 $\mu l \cdot l^{-1}$ ethylene, ACC synthase activity was detectable in AVG-treated apples (Fig. 2) and at 8.8 $\mu l \cdot l^{-1}$ saturation was almost achieved. No ACC synthase activity was detectable in AVG-treated apples ventilated with ethylene-free air.

Effects of ethylene on the content of ABA in ripening apples. Apart from C_2H_4 other hormones may be involved in the regulation of ripening of climacteric fruits (McGlasson et al. 1978). One possibility is ABA, since its concentration increases during the ripening of apple fruits (Rudnicki and Pieniazek 1970). When analysed in fruits treated with AVG and stored in ethylene-free air no increase in ABA concentration was detected, while in control fruits an almost 6- fold increase during 25 days at 20°C was observed (Fig. 3). Treating AVG-fruits continuously with 62 $\mu l \cdot l^{-1}$ C_2H_4 increased the concentration of ABA considerably. However, it is doubtful whether this increase in ABA is essential for ripening. Under the low partial pressure of O_2 in LPS ripening can be stimulated by the continuous addition of C_2H_4 (Bangerth 1975, Bufler

294

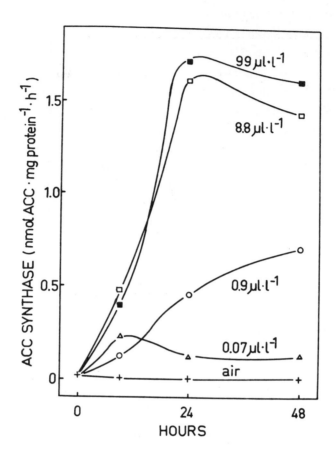

Fig. 2: Effect of different concentrations of ethylene on ACC-
synthase activity of AVG-treated postclimacteric apples (from
Bufler 1984).

and Bangerth 1981) without a simultaneous increase in ABA
(Bangerth 1980). ABA, therefore, does not seem to be a key
factor in the regulation of apple fruit ripening, at least under conditions of a low partial pressure of oxygen.

Similar experiments are possible, and are presently under
way, to investigate the role of IAA, which also changes considerably in concentration just before the onset of autocatalytic C_2H_4 production (Mousdale and Knee 1981). Since this
hormone shows close interactions with ethylene (Frenkel and
Dyck 1973, Lieberman et al. 1977, McGlasson et al. 1978)
fruits stored in LPS or treated with AVG should be an ideal

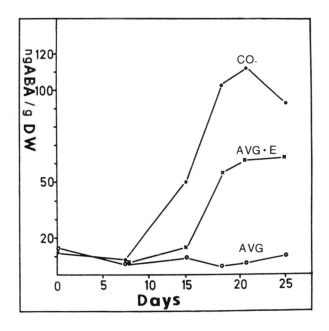

Fig. 3: Changes in ABA concentrations of control (Co.), AVG - and AVG + C_2H_4-treated fruits during storage at 20° (from Bangerth 1980).

material to study the significance of the changes in IAA concentration.

<u>The role of ethylene in controlling CO_2 production during the climacteric.</u> The correlation between the onset of ethylene production and the onset of climacteric respiration has long been known (Biale 1960). Ethylene and CO_2 production of apples stored under hypobaric conditions remain at a pre-climacteric level (Bangerth 1975, Bufler and Bangerth 1981). Ventilation of these fruits with ethylene, however, stimulates CO_2 production to a certain extent.

AVG-treated fruits are even more suitable for investigations on the effects of C_2H_4 on CO_2 production than fruits stored under hypobaric conditions, since C_2H_4 treatments can be conducted at normal atmospheric pressure, where oxygen is

not limiting. It was shown earlier that AVG-treated apples do
not exhibit a climacteric rise in respiration (Bangerth 1978).
When those fruits were treated with different concentrations
of ethylene, CO_2-production increased depending on the concentration of ethylene. This effect was saturated at ethylene
concentrations of somewhat above 5 $\mu l \cdot l^{-1}$ (Fig. 4).

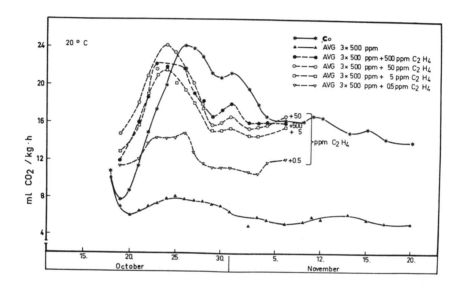

Fig. 4: Effect of different concentrations of ethylene on
the respiration of AVG-treated (3x500 mg/l) apple fruits at
20°C (from Halder-Doll 1982).

Removal of ethylene led to a sharp decline in CO_2 production
to an almost preclimacteric level (Bufler 1984). The repeated
removal and supply of ethylene caused a repeated decline and
increase in CO_2 production (Bangerth, unpubl.). Therefore, as
in non-climacteric fruits, CO_2-production of AVG-treated apples is dependent on ethylene concentration and requires the
continuous presence of ethylene.

Effects of ethylene on volatile production. Production of
volatiles in climacteric fruits increases drastically at the
onset of autocatalytic ethylene production (Drawert 1975).
Supressing ethylene biosynthesis and ripening by AVG inhibited
this increase in volatile production in apples at room temperature (Fig. 5) and also during a 4.5 months storage period at

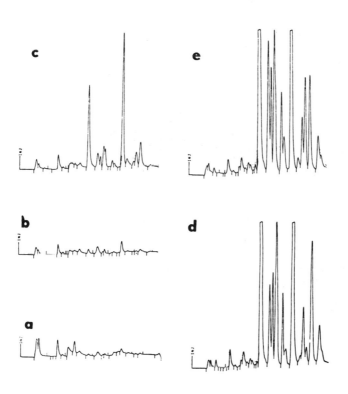

Fig. 5: Effect of ethylene (a=0; b=0.05; c=0.5; d=5; e=50µl·l^{-1})
on the production of volatiles of AVG-treated (2x500 mg/l)
fruits. Volatiles given off by the fruits were measured in the
outlet of the respirometer containers according to the method
of Streif (1981).

3°C (Halder-Doll 1982). However, if AVG-treated apples were
gassed at 20°C with different concentrations of ethylene, the
production of volatiles increased, dependent on the concentration of ethylene up to 50 µl·l^{-1} (Fig. 6). In contrast to

ACC synthase activity and respiration, volatile production declined more slowly when ethylene was removed Bangerth, unpubl.).

Apples stored under hypobaric conditions are kept in a preclimacteric stage and therefore do not produce signifikant amounts of volatiles during storage (Shatat et al. 1978). Ventilation of those fruits in LPS with C_2H_4 increased volatile production considerably. Also shelf-life conditions increased the production of volatiles. However, after 3 to 4 months storage under hypobaric conditions apples began to lose the capability for volatile production during their shelf-life, whereas autocatalytic C_2H_4 production is impaired only slightly. Loss of the capability for volatile production after prolonged storage of apples under controlled atmosphere (CA) conditions has also been reported (Patterson et al. 1974, Streif, person. comm.). Since ethylene and volatile production seem to be closely related, decreased volatile production after a prolonged storage period could be related to a decreased responsiveness of the tissue to ethylene.

Effects of ethylene on fruit softening. Stimulation of softening by ethylene has been reported for many fruits (McGlasson et al. 1978). When AVG-treated apples were ventilated at 20°C with ethylene, the rate of softening was enhanced by increasing concentrations of ethylene (Fig. 6). This effect was saturated at approximatly 50 $\mu l \cdot l^{-1}$. Without exogenous C_2H_4 softening of AVG-treated fruits was very slow.

Apples stored under hypobaric conditions did not soften appreciably during several months of storage (Fig. 7). When these fruits were treated continuously with ethylene they softened to about the same degree as cold-stored control fruits, provided the hormone was supplied from the beginning of the storage period onward. If the ethylene treatment was delayed for several months its effectiveness was considerably reduced (Fig. 8). Therefore, similar to respiration and volatile production, softening of apples after prolonged storage in LPS is disturbed.

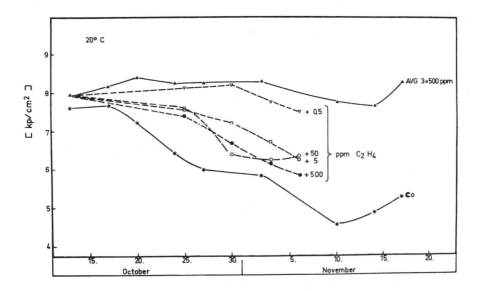

Fig. 6: Effect of different concentrations of ethylene on softening of AVG-treated (3x500 mg/l) apple fruits during storage at 20°C (from Halder-Doll 1982).

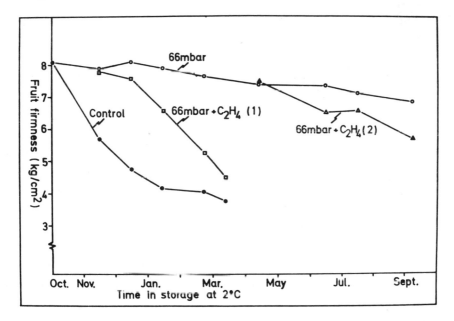

Fig. 7: Loss of fruit firmness during hypobaric (66 mbar ≙ 6.6 kPa) or cold (Control)-storage. Part of the LPS fruits were continuously gassed with ethylene, starting after harvest (66 mbar+C_2H_4 1) or at the 1st of May (66 mbar+C_2H_4 2).

CONCLUSIONS

The presented results suggest an intimate relationship between ethylene and various physiological changes during ripening of apple fruits. ACC synthase activity, respiration, volatile production and softening are all enhanced by similar concentrations of ethylene. The question arises therefore, whether these processes are regulated separately by ethylene or whether ripening is a closely integrated process.

In addition it was demonstrated that hypobaric storage and AVG are useful tools for investigations on the role of ethylene in fruit ripening. Nevertheless, possible side-effects of AVG and special features of hypobaric storage such as reduced partial pressures of oxygen and CO_2 must be kept in mind for interpretation of the presented results.

Acknowledgement: Financial support by the Deutsche Forschungsgemeinschaft in acknowledged.

REFERENCES

1. Adams DO, Yang SF. 1979. Ethylene biosynthesis: Identification of 1-aminocyclopropane-1-carboxylic acid as an intermediate in the conversion of methionine to ethylene. Proc.Nat.Acad. Sci. USA 76, 170-174
2. Bangerth F. 1975. The effect of ethylene on the physiology of ripening of apple fruits at hypobaric conditions. In Facteurs et Regulation de la Maturation des Fruits, Coll.Intern. C.N.R.S. No 238 Paris 1974, pp. 183-188
3. Bangerth F. 1978. The effect of a substituted amino acid on ethylene biosynthesis, respiration, ripening and preharvest drop of apple fruits. J.Amer.Soc.Hort.Sci. 103, 401-404
4. Bangerth F. 1980. Funktion der Abscisinsäure bei der Reife von Apfelfrüchten. Gartenbauwiss. 45, 224-228
5. Biale JB. 1960. Respiration of fruits. Role of ethylene and plant emanation in fruit respiration. Handb. Pflanzenphysiol. 12, 536-592
6. Bufler G. 1984. Ethylene-enhanced ACC synthase activity in ripening apples. Plant Physiol.(in press)
7. Bufler, G, Bangerth F. 1981. Enzymaktivitäten und Fruchtreife bei unterschiedlich gelagerten Apfelfrüchten. Gartenbauwiss. 46, 30-36
8. Bufler G, Bangerth F. 1983. Effects of propylene and oxygen on the ethylene-producing system of apples. Physiol. Plant. 58, 486-492
9. Burg SP, Burg EA. 1966. Fruit storage at subatmospheric pressures. Science 153, 314-315

10. Drawert F. 1975. Formation des aromes a differents stades de l'evolution du fruit; enzymes intervenant dans cette formation. In Facteurs et Regulation de la Maturation des Fruits, Coll. Intern. C.N.R.S. No 238, Paris 1974, pp. 309-318

11. Frenkel C, Dyck R. 1973. Auxin inhibition of ripening in Bartlett pears. Plant Physiol. 51, 6-9

12. Halder-Doll H. 1983. Auswirkungen des Ethylensynthesein-hibitors Aminoethoxyvinylglycin auf verschiedene prakti-sche und physiologische Parameter der Reife von Apfel-früchten. Diss. Hohenheim

13. Lieberman M. 1979. Biosynthesis and action of ethylene. Ann.Rev.Plant Physiol. 30, 533-591

14. Lieberman M. Baker JE. Sloger M. 1977. Influence of plant hormones on ethylene production in apple, tomato, and avocado slices during maturation and senescence. Plant Physiol. 60, 214-217

15. Lieberman M, Kunishi AT, Owens LT. 1975. Specific in-hibitors of ethylene production as retardants of the ripening process in fruits. In Facteurs et Regulation de la Maturation des Fruits. Coll.Intern. C.N.R.S. No 238, Paris 1974, pp. 161-169

16. McGlasson WB, Wade NL, Adato J. 1978. Phytohormones and fruit ripening. In Phytohormones and Related Compounds, Vol. II pp. 447-493

17. Mousdale DMA, Knee M. 1981. Indolyl-3-acetic acid and ethylene levels in ripening apple fruits. J.exp.Bot. 32, 753-758

18. Patterson BD, Hatfield SGS, Knee M. 1974. Residual effects of controlled atmosphere storage on the produc-tion of volatile compounds by two varieties of apples. J.Sci.Fd.Agric. 25, 843-849

19. Rudnicki R, Pieniazek J. 1970. The changes in concen-tration of abscisic acid (ABA) in developing and ripe apple fruits. Bull.Acad.Pol.Sci.Ser.Biol. 18, 577-580

20. Shatat F, Bangerth F, Neubeller J. 1978. Beeinflussung der Fruchtaromaproduktion durch drei verschiedene Lager-verfahren. Gartenbauwissenschaft 43, 214-222

21. Streif J. 1981. Vereinfachte Methode zur schnellen ga-schromatographischen Bestimmung von flüchtigen Aroma-stoffen. Gartenbauwiss. 46, 72-75

22. Yang SF. 1980. Regulation of ethylene biosynthesis. HortScience 15, 238-243.

POSSIBLE ROLE OF FRUIT CELL WALL OXIDATIVE ACTIVITY IN ETHYLENE EVOLUTION[1]

C. Frenkel and M. K. Mukai[2], Department of Horticulture and Forestry, Rutgers University, New Brunswick, New Jersey 08903

INTRODUCTION

The burst in ethylene evolution accompanying the respiratory upsurge in climacteric fruit has been extensively shown to accelerate the ripening process (3). The newly acquired knowledge of ethylene biosynthesis, particularly the finding that 1-amino cyclopropane-1-carboxylic acid (ACC) is the penultimate ethylene precursor (1) has been used to an advantage to further demonstrate the dependency of ripening on ethylene synthesis and action. Accordingly, depressed production of ACC, and subsequently, of ethylene evolution were also strongly inhibitory to the ripening process (23,29). Clearly, ethylene biosynthesis and action in climacteric fruit have a major regulatory role in ripening.

Considerable information is available regarding the synthesis of ACC. The mechanism of ethylene release from ACC is less well understood, however, although some guidelines indicate that the process is oxidative, requiring molecular oxygen, and is sensitive to uncouplers, temperature extremes, and other treatments (31). Voique et al (27) proposed that the requirement for oxygen may represent, in part, O_2 consumption during IAA oxidase dependent degradation of ACC. The authors suggest that although ACC is not directly metabolized by IAA oxidase the enzyme by-products, apparently active oxygen forms, can lead to the degradation of ACC and subsequently the release of ethylene. IAA oxidase activity was also implicated previously in the release of ethylene from other intermediates (13,17,18).

[1] New Jersey Agricultural Experiment Station, Publication No. D-12140-11-84 supported by State funds and by U.S. Hatch Act.
[2] Present Address: Universidate Federal de Vicosa, Departmento de Biologia Vegetal, 36570-Vicosa-MG, Brazil

Y. Fuchs and E. Chalutz (eds.) Ethylene: Biochemical, Physiological and Applied Aspects.
ISBN 90-247-2984-X. Printed in The Netherlands
©1984, Martinus Nijhoff/Dr W. Junk Publishers, The Hague.

Burg (4) outlined the dependency of ethylene synthesis on IAA and recent work established the auxin requirement for ACC production (32,34). The suggestion (26) that IAA may be required for the degradation of ACC, with the concommitent release of ethylene, represent an additional aspect of the phytohormone involvement in ethylene biosynthesis. This communication present data in support of this suggestion and furthermore indicate that in ripening fruit the process may entail cell surfaces activity.

MATERIALS AND METHODS

Whole pear (Pyrus communis var. Bartlett) at the mature green stage were used for testing ethylene evolution in fruit as influenced by auxins including IAA and 2,4-D, the auxin antagonist alpha-(p-chloropehnoxy) isobutyric acid (CPIBA), and the IAA-oxidase inhibitor 7-hydroxy-2,2-dimethyl-2,3-dihydrobenzofuran (HDDB). The compounds were administered in a mannitol carrier solution and applied by vacuum infiltration as outlined before (9). Ethylene evolution from whole fruit was measured as described previously (9,10).

Tomatoes fruit (Lycopersicon esculentum Mill.) variety ´Ramapo´ at the "breaking" stages were used to obtain cell wall preparation for the metabolic studies outlined below, essentially according to the method of Barnett (2). Fruit were peeled and 250 g outer pericarp tissue was frozen over dry ice. The tissue was then homogenized in 400 ml of 0.1 M potassium phosphate buffer, pH 7.4, containing 2 ml of n-octanol, 8 g ascorbic acid, and 8 g Polyclar AT as a PVP formulation for phenolics inactivation. The homogenization was carried out for 2 min in a Cuisinart Food Processor, followed by an additional 2 min homogenization at high speed in a Virtis homogenizer. The resulting slurry was filtered and rinsed through Miracloth filtering paper and washed with 16 to 20 l of double distilled and deionized water. The homogenization and rinsing were repeated. This procedure led to wall preparations virtually free of other cellular fractions including membrane residues (26). Wall preparations were used immediately or when necessary kept in sterile 20 mM citrate-phosphate buffer, pH 5.5 for incubation. For dry weight determinations, 25 ml wet volume of wall preparation were dehydrated at 50°C until no additional change in weight was observed. Wet wall preparations were used for testing the release of ethylene from ACC under the conditions outlined below.

The ethylene production reaction was carried out in 25ml serum-capped flasks. Head space samples were drawn by syringe and measured with a Model 5720 A Hewlett-Packard gas chromatograph with a flame ionization detector at 125°C, using a Poropak Q column in an oven temperature of 60°C. The complete reaction mixture included 1 ml of isolated cell wall, 30 mM potassium phosphate buffer, pH 7.6, 0.3 mM $MnCl_2$, 20 uM p-coumaric acid, 1 mM ACC with additions of NADH, IAA, or HDDB, as specified, in a total volume of 3 ml. Reaction time was 3 h.

RESULTS AND DISCUSSION

Stimulation of ethylene synthesis by auxins, as noted in other plant tissues (4), can also be demonstrated in fruit. In pears the applications of IAA stimulated the onset and the magnitude in the ethylene burst. A similar effect was obtained by 2,4-D, the synthetic auxin, although at supraoptimal concentrations, apparently toxic auxin level (9), the trend in ethylene production was reversed (Figure 1). In the climacteric fruit which normally display an upsurge in ethylene evolution the applied auxins led to an incremental addition in ethylene production whereas in other tissues displaying a nominal and low level in ethylene syntehsis the auxins initiate a major upsurge in the producton of ethylene. In the latter restricted supply of native auxins is presumably limiting to the process and this restriction can be relieved by an exogenous application of the compounds (32,34). It is reasonable to assume that in climacteric fruit, including pears, endogenous auxins, notably IAA, are in sufficient levels for stimulating a considerable ethylene synthesis. Hence, applied auxins, even at supraoptimal concentrations, led to only an additive effect.

To ascertain that native IAA in the fruit is available for the biosynthesis of ethylene, we attempted to antagonize the activity of the endogenous fruit auxin. Previously, alpha (p-chlorophenoxy) isobutyric acid (CPIBA) was used as a formulation to inhibit IAA dependent processes, including the extension of _Avena_ coleoptiles

(21) or the curviture of citrus flower petals (11). Provided that in climacteric fruit auxin activity is sufficient to stimulate ethylene production the compound could lead to the inhibition of auxin dependent ethylene synthesis (10). The results (Figure 2) show that to be the case; the application of CPIBA led to progressive deferal in the onset and reduction in the magnitude of ethylene evolution as the concentrations of the compound increased. These results are appropriately the inverse of the auxin stimulated ethylene evolution (Figure 1). These data further support the suggestion that substantial auxin activity in ripening climacteric fruit may engage and stimulate the ethylene biosynthetic pathway, although proof for this suggestion requires the measurement of auxins turnover as related to the onset of ethylene evolution in climacteric fruit.

IAA-oxidase is one of the metabolic systems showing affinity toward IAA, and therefore, may be used as a test system to examine the mode of auxin involvement in ethylene biosynthesis, especially since the enzyme was implicated in the release of ethylene from ACC (24,27). To examine this possibility whole pears were treated with an inhibitor of IAA oxidase, the carbofuran HDDB. The compound inhibits the formation of reactive enzyme forms (14,15) and, thereby, reactions which are catalysed by IAA oxidase. Figure 3 shows that HDDB depressed the evolution of ethylene in whole fruit, and although not complete, the inhibition of the process was concentration dependent. These results are in keeping with the suggestion that ethylene evolution is, at least in part, dependent on the oxidative degradation of IAA as catalyzed by the activity of IAA oxidase and that interference with the process is accordingly inhibitory to ethylene evolution.

To directly examine this hypothesis IAA-oxidase was extracted from tomato fruit and tested on one hand for the ability to catalyze the release of ethylene from ACC and on the other to respond to the conditions which restrict ethylene evolution in the intact fruit. Purified cell wall preparations from tomato fruit were used as a source of IAA oxidase activity since the organelle is a major site for the enzyme activity (28). Figure 4 shows that purified preparation of the cell wall show only limited ability for releasing ethylene from ACC. However, a stepwise increase in the IAA concentration in the reaction medium led to progressively higher level of released ethylene. Conversely, in a complete reaction medium containing IAA, the IAA oxidase inhibitor, HDDB, depressed the release of ethylene from ACC, in proportion to the inhbitor concentration (Figure 5). These results showing that ethylene evolution is stimulated by IAA and conversely is restricting by inhibiting IAA degradation are in agreement with the effect of the test compounds in whole fruit. It appears, therefore, that the turnover of auxins in fruit as catalyzed by IAA oxidase activity may lead to the release of ethylene from ACC as previously suggested (24,27).

The mode of the enzyme action suggested by background studies may be indirect, probably by furnishing active oxygen species (27) and possibly other active intermediates resulting form the oxidative degradation of IAA or other enzyme substrate (30). Accordingly when propyll gallate, an antioxidant, is introduced into the reaction median, the release of ethylene was strongly inhibited (Table I). In this respect the function of IAA oxidase is analogous to the action of other oxidative systems including isolated chloroplast (6) and microsomal fractions (19,20,22) or non enzymatic systems (16) which can lead to release of ethylene from precursor compounds, including ACC, by generating active oxygen species.

There are some difficulties with this concept, however. Other data infer formation of ACC-enzyme complex (19) as shown by stereospecific binding of ACC (12) and also competitive displacement of ACC by analogues (25). In these ethylene forming systems ACC saturation at micro-mole concentration and high efficiency are obtained in the conversion of ACC to ethylene. Moreover, the process may be dependent on ATP availability (33) and subject to other controls (31), and therefore, a coordinated process. By comparison, the view espoused in this presentation suggest that ACC degradation results from and is coincidental to the production of active reaction products of IAA oxidase. Secondly, the process involving the oxidase action appears to be unsaturable with regard to ACC and is characterized by fairly low efficiency in the conversion of the compound to ethylene. Thus, the action of oxidases, including IAA oxidase, may approach reaction conditions in which the release of ethylene from ACC by active oxygen species are similar to non-enzymatic systems (16). In addition, other oxygen utilizing systems in plant tissues may also be active in the process, and accordingly it is possible to demonstrate ethylene evolution in the course of NADH oxidation in chloroplasts (6) or from ACC during oxygen utilization by microsomal fractions (19,20,22). It is difficult, therefore, to ascertain which oxidase system may be differentially accessible to ACC or whether the process is generally a by-product of the tissue oxidative activity. These features are incongruous with the concept that the release of ethylene from ACC is a highly coordinated process.

These views need not be mutually exclusive, however. Ethylene evolution was shown in fruit discs to be a cell surface phenomenon (8), and furthermore, may be associated with cell wall modifications, for example, ethylene evolution during cell wall changes in host tissues challenged by fungal elicitor (7), or in tobacco tissue following the application of wall degrading enzyme preparations (5). These data are in agreement with the results (not shown) indicating that the release of ethylene from ACC by fruit wall preparations is, in part, dependent on the development of the wall oxidative activity as walls undergo dissolution during ripening. It is likely that wall changes, as occurring in normally ripening fruit, tissues subjected to fungal elicitors, or other wall modifying treatments, are asociated with the increase in the tissue oxidative activity. In turn, a complementary system in the cell wall-membrane interface may represent the ACC complexing site. This scheme could account for the observation that the ethylene releasing activity in fruit is cell surfaces phenomenon, the requirements for a highly coordinated process and the observation that cell wall oxidative activity, notably IAA oxidase, function could lead to ACC degradation.

A clearly defined ethylene releasing system will undoubtedly help to settle this question and at present the explanation for the involvement of the fruit IAA oxidase or other oxidases in the release of ethylene from ACC may be regarded as tentative.

Table I.   Ethylene release from ACC by cell wall preparations, from ripening "Ramapo" tomato fruit, as influenced by propyl gallate. The reaction medium consisted of 35.9 g equivalents of dry weight wall material, 30 mM phosphate buffer, pH 7.6, 1 mM IAA, 1 mM ACC, 0.3 mM $MnCl_2$, and 0.02 mM p-conmaric acid. Ethylene was allowed to evolve for 3 h in a closed vessel, as outlined in the methodology.

Additions	ethylene, nl
none (control)	5.37 + 0.09
propyl gallate, 10 uM	0.15 + 0.02
propyl gallate, 100 uM	0.13 + 0.02

REFERENCES
1.   Adams D O, Yang S F. 1979. Ethylene biosynthesis: identification of 1-amino-cyclopropane-1-carboxylic acid as an intermediate in the conversion of methionine to ethylene. Proc Natl Acad SC: USA 76, 170-174.
2.   Barnett N M. 1974. Release of peroxidase from soybean hypocotyl cells by Sclerotium rolfsii culture filtrates. Can J Bot 52, 265-271.
3.   Burg S P, Burg E A. 1965. Ethylene action and the ripening of fruit. Science 148, 1190-1196.
4.   Burg S P, Clogett C O. 1967. Conversion of methionine to ethylene in vegetative tissue and fruits. Biochem Biophys Res Commun 27, 125-130.
5.   Chaluz E, Mattoo A K, Solomos T, Anderson J D. 1984. Enhancement of cellulysin-induced ethylene production by tobacco leaf discs. Plant Physiol 74, 99-103.
6.   Elstner E F, Konz J R. 1974. Light dependent ethylene production by isolated chloroplasts. FEBS-Lett 45, 18-21.
7.   Esquerre Tugaye M T, Mazau D, Toppan A. 1983. Hydroxyproline-rich glycoproteins i cell wall of diseased plants as a defense mechanism. In: Post-harvest Physiology and Crop Preservation. M Lieberman, ed, Plenum Press, New York, London, pp 287-298.
8.   Forney C F, Arteca R N, Walner S J. 1982. Effect of amino and sulfhydryl reactive reagents on respiration an ethylene production in tomato and apple fruit discs. Physiol Plant 54, 329-332
9.   Frenkel C, Dyck R. 1973. Auxin inhibition of ripening in Bartlett pears. Plant Physiol 51, 6-9.
10. Frenkel C, Haard N F. 1973. Initiation of ripening in Bartlett pear with an antiauxin alpha (p-chlorophenoxy) isobutyric acid. Plant Physiol 52, 380-384.
11. Goldschmidt E E, Monselise S P. 1966. Citrus petal-bioassay based on the indolyl-3-acetic effect on flower opening. Nature 212, 1064-1065.

12. Hoffman N E, Yang S F, Ichihara A, Sakamara S. 1982. Stereospecific conversion of 1-amino-cyclopropane-1-carboxylic acid to ethylene by plant tissues. Plant Physiol 70, 195–199.

13. Ku H S, Yang S F, Pratt H R. 1969. Ethylene formation from alpha-kato-gamma-methyl butyrate by tomato fruit extracts. Phytochemistry 8, 567–573.

14. Lee T T. 1977. Role of phenolic inhibitors in peroxidase-mediated degradation of indole-3-acetic acid. Plant Physiol 59, 372–375.

15. Lee T T, Chapman R A. 1977. Inhibition of enzymic oxidation of indole-3-acetic acid by metabolites of the insecticide carbofuran. Phytochem 16, 35–39.

16. Legge R L, Thompson J E, Baker J E. 1982. Free radical-mediated formation of ethylene from 1-amino-cyclopropane-1-carboxylic acid: a spin trap study. Plant Cell Physiol 23, 171–177.

17. Mapson L W, Wardale D A. 1971. Enzymes involved in the synthesis of ethylene from methionine, or its derivatives in tomatoes. Phytochemistry 10, 29–39.

18. Mapson L W, Wardale D A. 1972. Role of indole-3-acetic acid in the formation of ethylene from 4-methylmercapto-2-oxo butryic acid by peroxidase. Phytochemistry 11, 1371–1387.

19. Mattoo A K, Achilea O, Fuchs Y, Chalaz E. 1982. Membrane association and some characteristics of the ethylene forming enzyme from ethiolated pea seedlings. Biochem Biophys Res Commun 105, 271–278.

20. Mayak S, Legge R L, Thomspon J. E. 1981. Ethylene formation from 1-amino-cyclopropane-1-carboxylic acid by microsomal membranes from senescing carnation flowers. Planta 153, 49–55.

21. McRae D H, Bonner J. 1953. Chemical structure and antiauxin activity. Physiol Plant 6, 485–510.

22. McRae D G, Baker J E, Thompson J E. 1982. Evidence for the involvement of the superoxide radical in the conversion of 1-amino-cyclopropane- 1-carboxylic acid to ethylene by pea microsomal membranes. Plant Cell Physiol 23, 375–383.

23. Ness J P, Romani R J. 1980. Effect of amino ethoxy vinyl glycine and counter-effect of ethylene on ripening of Bartlett pear fruit. Plant Physiol 65, 372–376.

24. Rohwer F, Mader M. 1981. The role of peroxidase in ethylene formation from 1-aminocyclopropane-1-carboxylic acid. Z Pflanzenphysiol Bd 194, 363–372.

25. Satoh S, Esashi Y. 1980. Alpha-aminobutyric acid: A probable competitive inhibitor of conversion of 1-aminocycloporpane-1-carboxylic acid to ethylene. Plant Cell Physiol 21, 939–949.

26. Strand L L, Mussell H. 1975. Solubization of peroxidase activity from cotton cell walls by endopolygalacturonases. Phytopothal 65, 830–831.

27. Vioque A, Albi M A, Vioque B. 1981. Role of IAA-oxidase in the formation of ethylene form 1-aminocyclopropane-1-carboxylic acid. Phytochem 7, 1473–1475.

28. Waldrum J D, Davis E. 1981. Subcellular localization of IAA oxidase in peas. Plant Physiol 68, 1303-1307.

29. Wang C Y, Mellenthin W M. 1977. Effect of amino ethoxy analogue of rhizobitoxine on ripening of pears. Plant Physiol 59, 548-549.

30. Yamazaki I, Yokota K-N, Nakajima R. 1977. Reactions of free radicals with molecular oxygen. In: Biochemical and medical aspects of active oxygen. O Hayaishi, K Asada, eds, Univ Tokyo Press, Tokyo, pp 91-100.

31. Yang S F, Adams D O, Lizada C, Yu Y B, Bradford K J, Cameron A C, Hoffman N E. 1980. Mechanisms and regulation of ethylene biosynthesis. In: F Skoog, ed, Proc 10th International Conference on Plant Growth Substances Springer-Verlag, Berlin, pp 219-229.

32. Yu Y B, Adams D O, Yang S F. 1979. Regulation of auxin-induced ethylene production in mung bean hypocotyls: Role of 1-amino-cyclopropane-1-carboxylic acid. Plant Physiol 63, 589-590.

33. Yu Y B, Adams D O, Yang S F. 1980. Inhibition of ethylene production by 2,4-Dinitrophenol and high temperature. Plant Physiol 66, 286-290.

34. Yu Y B, Yang S F. 1979. Auxin-induced ethylene production and its inhibition by amino ethoxy vinyl glycine and cobalt ions. Plant Physiol 64, 1074-1077.

ILLUSTRATIONS

Figure 1.   Stimulation of ethylene evolution in ripening Bartlett
             pear by IAA (a) and 2,4-D (b). The auxin
             concentrations, in the applied test solutions, were zero
             (mannitol control) (O); 0.01 mM (●); 0.1 mM (□);
             and 1.0 mM (■).

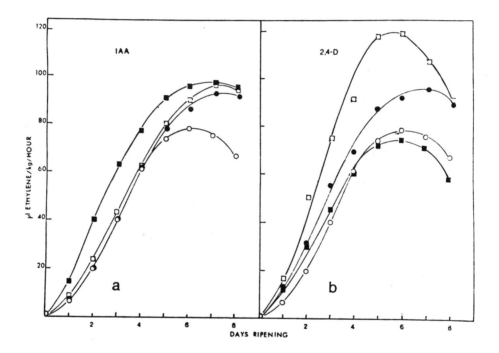

Figure 2.   Ethylene evolution in ripening Bartlett pears as
             influenced by the auxin antagonist CPIBA.   The
             concentrations of CPIBA in the applied test solutions
             were zero (mannitol control) (O); 0.02 mM (▲);
             0.2 mM (■); and 2.0 mM (◆).

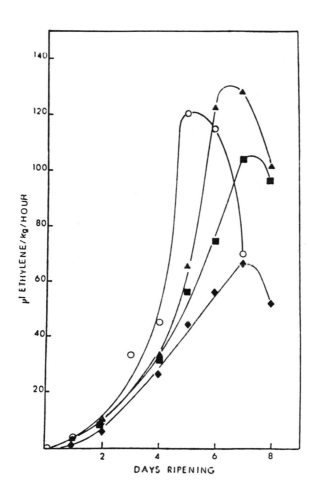

Figure 3.   Evolution of ethylene in ripening Bartlett pears as influenced by HDDB, the IAA oxidase inhibitor. The concentrations of HDDB in the applied test solutions were zero (mannitol control) (●); 0.01 mM (○); and 1.0 mM (▲).

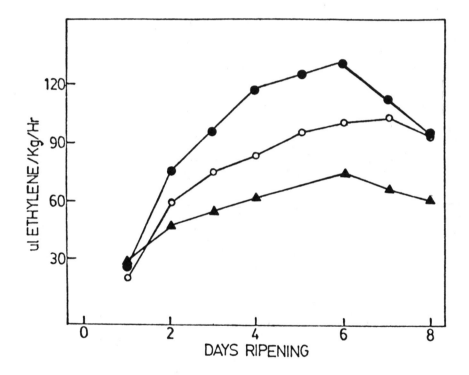

Figure 4.   Influence of different IAA concentrations on release of
ethylene from ACC by cell wall preparations from
ripening "Ramapo" tomato fruit. The assay conditions
were outlined in the methodology. The wall material
represented 30.2 mg equivalent of dry weight.

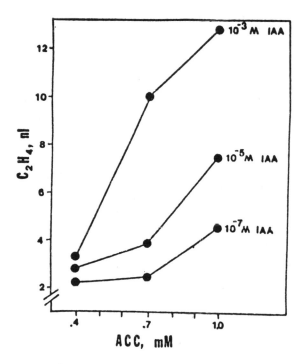

Figure 5.   Influence of HDDB, the IAA oxidase inhibitor, on
ethylene release from ACC by cell wall preparations from
ripening "Ramapo" tomato fruit. The IAA concentrations
in all treatments were kept at 1.0 mM.   The wall
material represented 42.7 mg equivalent of dry weight.

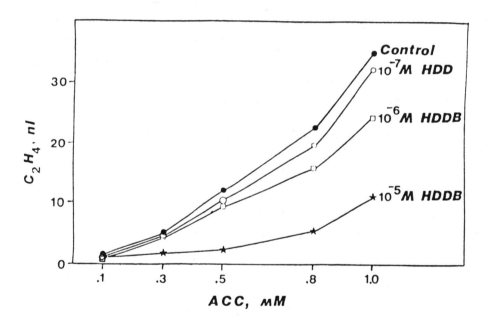

COMPARTMENTATION OF AMINO ACIDS IN TOMATO FRUIT PERICARP TISSUE

J. E. BAKER and R. A. SAFTNER
USDA, ARS, Plant Hormone Laboratory
Beltsville, Md. U.S.A.20705

INTRODUCTION

During ripening of tomato fruit, there is a progressive decline in labeled amino acid incorporation into proteins of pericarp tissue (3). The specific radioactivity of ethylene produced from [14C]-methionine is also lower in ripe tissue than in green tissue and ethylene production in ripe tissue is relatively insensitive to aminoethoxyvinylglycine (AVG) (2). Also, the rate of ethylene production by tomato pericarp tissues is, under certain conditions, less than would be expected on the basis of ACC content (7). These phenomena could be due to changing uptake or compartmentation patterns of amino acids as the fruit ripens. A decompartmentalized methionine pool might dilute added labeled methionine resulting in lower incorporation of label into proteins and ethylene. A knowledge of amino acid uptake and compartmentation characteristics in tissue at various stages of ripening is important in understanding these observations, and the regulation of 1-aminocyclopropane-1 carboxylate (ACC) production and conversion to ethylene in tomato fruit.

PROCEDURE

Material and methods

Tomato fruit. Tomato (Lycopersicon esculentum Mill, cv Rutgers) fruit were grown at the Beltsville Agricultural Research Center. For the experiments on uptake and incorporation of [14C]-methionine, cv Sunny fruit were obtained from a local wholesale market. For uptake and incorporation of [14C]-methionine, slices (2 mm) were cut with a meat slicer, and the outer pericarp was sectioned with a razor blade into blocks ca 5 mm^2 x 2 mm. For the α-aminoisobutyric acid compartmentation studies, the skin was removed and the slices were 1 mm in thickness.

Incorporation of label from L-[3,4-14C]-methionine into protein and ethylene. Pericarp slices were washed in 0.4 M sorbitol-0.01 M 2-(N-morpholino) ethanesulfonic acid (MES) buffer, pH 6.7, and blotted with Kaydry[1] plastic-backed bench paper. Tissue (2 g)

---

[1]Mention of a company name or trademark does not constitute endorsement by the U.S. Department of Agriculture over others of a similar nature not mentioned.

Y. Fuchs and E. Chalutz (eds.) Ethylene: Biochemical, Physiological and Applied Aspects.
ISBN 90-247-2984-X. Printed in The Netherlands
©1984, Martinus Nijhoff/Dr W. Junk Publishers, The Hague.

was incubated in 3 ml of the above buffer containing 0.5 μCi/ml L-[3, 4-^{14}C]-methionine, sp. act. 57 μCi/μmole, and incubated in 25 ml Erlenmeyer flasks for 3-4 h at 25°C. At the end of this period, the incubation medium was removed and fresh unlabeled medium was rapidly pipetted into the flasks and removed within 5 sec. The tissue was ground in 10% trichloroacetic acid (TCA) and total radioactivity and TCA-precipitable radioactivity (11) was determined. Radioactive ethylene was measured by the method of Aharoni et al (1).

Analysis of α-[3-^{14}C]-aminoisobutyric acid compartmentation: The localization of α-AIB within the tissue was determined using the compartmental efflux method (see reference 9 for additional information related to this method). Sets of tissue slices were blotted and then equilibrated (1 g/3 ml) for 120 min in an aerated solution (0.2 M sorbitol, 0.02 M sucrose, 0.001 M CaCl$_2$, 0.001 M MES buffer adjusted to pH 5.5 with 1.0 N KOH). The solution was replaced after 30, 60, and 90 min. After equilibration, 1 g sets of tissue slices were incubated in 3 ml of aerated incubation medium containing 25 or 200 μM α-AIB, 0.02 M sucrose, 0.001 M CaCl$_2$, 0.01 M MES buffer adjusted to pH 5.5 with 1.0 N KOH, and sufficient sorbitol to make the osmolality of the incubation medium identical to that of the equilibration solution. The specific activity of the incubation medium was 19.9 and 0.3 μCi/μmole α-[3-^{14}C]-AIB at 25 and 200 μM α-AIB, respectively. Incubation was at room temperature (approximately 21°C) for periods of 2 to 21 h during which time net influx remained linear. After incubation in ^{14}C-α-AIB solutions, each set of tissue slices was washed with 2 ml samples of unlabeled incubation medium at various time intervals. After a 5 h wash period, each set of tissue slices was then placed in a Soxhlet extraction apparatus and α-AIB extracted for 3 h with boiling 80% ethanol. This process removed more than 99% of the radioactivity from the tissue. Ethanol was removed by boiling, and the residue dissolved in 2 ml of unlabeled incubation medium. Radioactivity in tissue wash and extraction samples was determined by liquid scintillation spectroscopy, and efficiency was determined by Beckman Instrument's quench monitoring method. All samples were run in triplicate for each experiment. It was determined that the recovered label chromatographed in the position of authentic α-aminoisobutyric acid, and that a negligible amount of radioactivity was present in the TCA-insoluble fraction after incubation of the tissue with labeled AIB. Also, it was determined that chloramphenicol at 50 μg/ml in the incubation medium had no effect on uptake and kinetics of exchange of AIB, indicating that microbial contamination was not a problem in these experiments.

The free space volume of tomato pericarp tissue, at various stages of development, was determined by the method of Parr and Edelman (12). Slices were incubated with [^{14}C]-sorbitol or α-[^{14}C]-AIB for 1 h and the activity in the free space was quantified by compartmental analysis. From the specific activity of the external medium and the ^{14}C in the free space, the volume of the free space was estimated. The volumes of the cytoplasm and vacuole were estimated from light microscopic measurements of tomato pericarp parenchymal cells.

Chemicals

L-[3,4-^{14}C]-methionine, (57 μCi/μmole) was obtained from Research Products International. Uniformly labeled [^{14}C]-sorbitol (200 μCi/μmole) and α-[3-^{14}C] AIB (19.9 μCi/μmole) were obtained from New England Nuclear. Other chemicals were reagent grade from major suppliers.

RESULTS AND DISCUSSION

Uptake and incorporation of [^{14}C]-methionine by tomato pericarp slices

In Table 1 it is shown that both uptake and incorporation of [^{14}C]-methionine into TCA-insoluble materials decline with ripening. The decline of incorporation (> 19 fold) is considerably greater than that of uptake (< 3 fold) and this is reflected in a declining incorporation/uptake ratio. The values for uptake include surface adsorbed label, and do not reflect metabolism of methionine to CO_2 and ethylene. The values can only be regarded as rough estimates of methionine uptake, and do not give any information about compartmentation of methionine or its metabolites within the cell.

Table 1. Uptake and incorporation of L-[3,4-^{14}C]-methionine into TCA-insoluble materials by tomato pericarp tissue

Stage of fruit	Uptake	Incorporation (TCA-insoluble fraction)	Ratio of incorporation to uptake
	dpm x 10^{-4}	dpm x 10^{-4}	
Mature green	245.7	115.3	0.47
Pink	143.0	32.4	0.23
Red	93.3	6.0	0.06

Conversion of ^{14}C-methionine to ethylene

The specific activity of [^{14}C]-ethylene produced from L-[3,4-^{14}C]-methionine is about 3.5 times lower in red than in mature green tomato pericarp tissue (Table 2). Thus, the specific activity of ethylene does not decrease to the same extent as does incorporation of label into proteins. If both processes occur in the soluble cytoplasm, this would indicate that the decrease in [^{14}C]-methionine incorporation into proteins is not entirely due to transport or compartmental changes in red tissue.

Table 2. [^{14}C]-ethylene production from L-[3,4-^{14}C]-methionine by tomato pericarp tissue at various developmental stages

Stage of fruit	C_2H_4 nmoles/g x h	$^{14}C_2H_4$ dpm/g x h	Specific activity dpm/nmole
Immature green	0.38	3652	9611
Mature green	0.63	7948	12616
Breaker	2.20	18811	8550
Red	1.1	3990	3627

Estimation of α-AIB compartmentation and fluxes

The observed lower incorporation of labeled methionine into protein and the decrease in specific activity of ethylene produced from ^{14}C-methinine during tomato pericarp development could be due to changes in membrane flux and/or cellular compartmentation patterns as the fruit ripens. However, transport and compartmentation studies of methionine, ACC(the ethylene precursor), and other neutral amino acids are complicated by metabolic changes in the substrate that are, at best, only indirectly related to the transport and compartmentation processes. Neutral amino acid transport/compartmentation can be separated from subsequent metabolism and incorporation into proteins by using the non-toxic neutral amino acid α-AIB, that is transported/compartmentalized, but not metabolized by plant tissues, including tomato pericarp tissues. α-AIB is also a structural analog of ACC and can inhibit ACC conversion to ethylene both in vivo and in vitro, but only at high (mM) concentrations (16). In the present study, α-AIB was used to investigate neutral amino acid transport/compartmentation patterns during development in tomato pericarp.

Efflux of labeled α-AIB from tomato pericarp was measured over a period of 5 h using triplicate samples of tissue preincubated for various periods of time in either 25 or 200 μM labeled α-AIB. From counts of radioactivity remaining in the tissue at the end of the elution, and from counts in the washing solutions, an efflux curve was constructed for each sample (Fig. 1). This curve is a compound curve made up of first-order rate losses from the various compartments within the tissue (10). Regression analyses yield a series of straight lines each representing a particular compartment. The contribution of the slowest compartment is assumed to be the vacuole and is represented by the final linear phase in Fig. 1. The resulting line has a slope which is the rate constant for passive efflux from the vacuole and the intercept gives the amount of radioactivity in the vacuole at the beginning of elution. Since α-AIB is not a natural constituent of tomato fruit, the specific activity of α-AIB within the tissue is the same as the specific activity of the α-AIB applied to the tissue. Therefore, the content of α-AIB in the vacuole at the beginning of elution can be

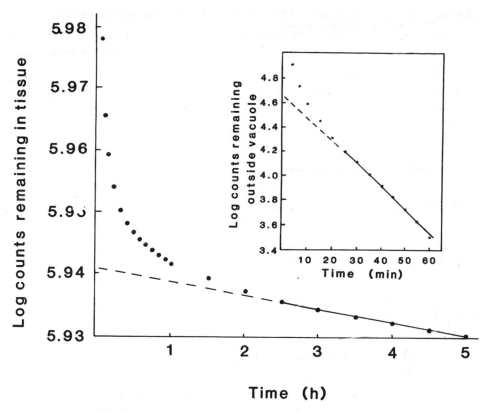

Fig. 1 Loss of [^{14}C]-α-aminoisobutyric acid (α-AIB) from mature green tomato fruit pericarp tissue to unlabeled washing medium. Tissue was loaded for 8 h in medium containing 200 μM α-AIB, sp act. 0.3 μCi/μmole.

calculated. Subtraction of the vacuolar component from the activity of the total tissue at each time interval yields another compound efflux curve representing passive efflux from all tissue compartments other than the vacuole with the final linear phase being attributed to the cytoplasmic compartment (Fig. 1 insert). The slope of this second line is the rate constant for passive efflux from the cytoplasm, and the intercept of the line gives the initial amount of radioactivity in the cytoplasm. Dividing the intercept radioactivity by the specific activity of α-AIB gives the α-AIB content in the cytoplasm at the beginning of elution. Further regression analyses yield two additional compartments, which have been related to the cell-wall free space and the surface film of α-AIB solution on the tissue (17).

The passive efflux of α-AIB from the cytoplasm must cross only the plasmalemma. Since the fraction of the tissue volume occupied by the cytoplasm (Table 3) and the volume/surface area ratio (17) in tomato pericarp remain relatively constant during growth and development, the cytoplasmic rate constant can be used as a relative measure of the passive permeability of the plasmalemma of pericarp cells in tomato fruit at various developmental stages (see reference 17 for more details). The interpretation of the vacuolar rate constant is more complicated depending on whether or not vacuolar α-AIB effluxes through the cytoplasmic compartment. If vacuolar α-AIB effluxes through the cytoplasm, vacuolar efflux to the washing solution must cross both the tonoplast and the plasmalemma. In such a case, and since the fraction of the tissue volume occupied by the vacuole (Table 3) and the volume/surface area ratio (17) does not dramatically change in tomato pericarp during development, the vacuolar rate constant can be used as a measure of the combined passive permeabilities of the tonoplast and plasmalemma. This interpretation of vacuolar rate

Table 3. Compartment volumes in tomato pericarp tissue at various stages of fruit development.

Stage of fruit	Vacuole		Cytoplasm		Free Space	
	%	µl	%	µl	%	µl
Immature green	78.8	867	3.9	42.5	6.8	74.5
Mature green	81.7	899	3.7	40.2	7.2	79.0
Breaker	81.5	896	3.5	38.6	6.9	76.0
Pink	82.2	904	3.3	36.7	8.7	95.5
Red	81.9	901	3.3	36.2	8.8	96.4

constants was used by Vickery and Bruinsma (17) to study K^+ compartmentation in tomato pericarp, and is the interpretation we applied in this study.

From the graphical analysis of the results (Fig. 1), the efflux rate constants (k) and the half-time for loss ($t_{1/2}$=ln 2/k, reference 10) of α-AIB from each cellular compartment can be estimated. The results are presented in Table 4. The rate constants and half-time values for the cell-wall free space, and surface-film compartments did not noticeably change during fruit development. The average half-time for passive efflux of α-AIB were 2.1 and 0.9 min for the free space, and surface film compartments respectively. The average rates constants were 0.33 and 0.75 min^{-1} respectively. In general, these values agree with data from washouts of organic molecules and inorganic ions from other plant tissues (15) and particularly well agree with kinetic values for K^+ efflux in tomato pericarp (17). The rate constant and half-time values for the cytoplasmic compartment did not noticeably change from the immature green to the breaker stage of development. During

Table 4. Rate constants and $t_{1/2}$ values for α-AIB efflux from tomato pericarp compartments.

Stage of fruit	Compartment	Rate Constant	Halftime
		min^{-1}	min
Immature green	Vacuole	6.47×10^{-5}	10,700
	Cytoplasm	2.50×10^{-2}	27.7
	Free space	0.35	2.0
	Surface film	0.77	0.9
Mature green	Vacuole	8.77×10^{-5}	7,900
	Cytoplasm	2.56×10^{-2}	27.1
	Free Space	0.32	2.1
	Surface film	0.73	1.0
Breaker	Vacuole	8.55×10^{-5}	8,100
	Cytoplasm	2.64×10^{-2}	26.2
	Free space	0.34	2.0
	Surface film	0.75	0.9
Pink	Vacuole	1.10×10^{-4}	6,300
	Cytoplasm	4.15×10^{-2}	17.0
	Free space	0.32	2.2
	Surface film	0.76	0.9
Red	Vacuole	2.57×10^{-4}	2,700
	Cytoplasm	7.36×10^{-2}	9.4
	Free space	0.33	2.1
	Surface film	0.75	0.9

this developmental period, the cytoplasmic rate constant averaged 2.57×10^{-2} min^{-1} and the half time 27.0 min. However, during the later stages of fruit ripening, the cytoplasmic rate constant increased greatly and the half-time correspondingly declined (Table 4). Such changes may indicate an increasing passive permeability of the plasmalemma for α-AIB during fruit ripening. In one experiment with high α-AIB concentration (200 μM), there was no indication for an increased passive permeability of the plasmalemma for α-AIB. The permeability characteristics for the plasmalemma just mentioned apply only to passive, independent fluxes of α-AIB. Changes in plasmalemma permeability resulting from active or non-independent fluxes during fruit development are still possible. The data in Table 4 also show that the rate constants for α-AIB loss from the vacuole increased from 6.5×10^{-5} to 2.6×10^{-4} min^{-1} and the half-time values decreased from 10,700 to 2,700 min. Since the passive permeability of the plasmalemma apparently did not change during the early stage of fruit ripening, the large changes in rate constant and half-time values for the vacuolar compartment between the immature green and the breaker stage of development indicate that the passive permeability of the tonoplast for α-AIB increases dramatically during this period of development. The passive permeability of the tonoplast for α-AIB may continue to increase during later stages of fruit ripening but this cannot be ascertained because the passive permeability of the plasmalemma also increases during this period, a process which could influence the passive efflux of α-AIB from the vacuole. In any case, the overall passive efflux of α-AIB from the vacuole increases throughout fruit development. These results are in constrast to those of Vickery and Bruinsma (17) who found no change in the passive permeabilities of the tonoplast or plasmalemma for K^+ions during growth and development in tomato pericarp. If the passive permeabilities of the tonoplast and plasmalemma change during fruit development for α-AIB but not for K^+ ions, the permeability changes are rather specific in nature, and are not the result of gross structural changes in the membranes.

From tissue incubations of 2 to 21 h, the net influx rate into the vacuole was approximately linear for each stage of development (Saftner and Baker, unpublished results). However, the net influx rate into the vacuole declined during fruit growth and development. This may be due to a relatively greater influx into and/or relatively lesser efflux from the vacuoles of cells at the earlier stages of development. If the tonoplast of pericarp cells becomes more permeable to α-AIB as fruit develop, part, if not all, of the differences between tissue stages in net influx rates into the vacuole could be explained as being due to increased tonoplast permeability. In addition, more than 95% of the net influx of α-AIB into the vacuole is inhibited by treatment with metabolic uncouplers such as CCCP and by anaerobic treatment (Saftner and Baker, unpublished results).

The α-AIB content in each cellular compartment for each developmental stage can be determined from the graphical analyses (as in Fig. 1). Dividing the α-AIB content for each compartment by its volume (Table 3 values multiplied by the total tissue volume of about 1100 µl per g) gives the concentration of α-AIB for each compartment at each developmental stage. The results are reported in Table 5 for tissues preincubated with 25 µM α-AIB for 17 h and with 200 µM α-AIB for 4 h. When tomato pericarp tissues are exposed to a high (200 µM) or low (25 µM) concentration of α-AIB, the vacuolar concentration of α-AIB at the beginning of elution is highest in immature green tissue and steadily declines at later stages of fruit development. The cytoplasmic concentration increases somewhat during fruit development when tissue is preincubated in 200 µM α-AIB, but greatly declines in the latter stages of fruit development, when incubations are with 25 µM α-AIB. The α-AIB concentration in the cell-wall free space remained approximately the same as in the

Table 5. Micromolar concentrations of α-AIB in tomato pericarp compartments after incubation in 200 μM α-AIB (A) for 4 h, and 25 μM α-AIB (B) for 17 h.

Stage of fruit	Vacuole	Cytoplasm	Free space
		A	
Immature green	140	104	150
Mature green	54	168	178
Breaker	53	194	178
Pink	38	241	183
Red	11	272	192
		B	
Immature green	17.6	319	15.4
Mature green	10.5	185	19.4
Breaker	10.2	189	19.5
Pink	4.6	71	22.7
Red	3.2	51	23.4

incubation solution. The α-AIB concentration in the external solution declined during incubation as α-AIB influxed into the tissue. Since net influx was greater in early developmental stages, the external α-AIB concentration declined faster in these stages. Transport of α-AIB into the vacuole is a metabolically-dependent process in that uncouplers such as CCCP and anaerobic treatments inhibited net influx more than 95%. In addition, all tissue stages except red tissue were found capable of accumulating α-AIB against its concentration gradient into the vacuole if a suifficient period of incubation was allowed (Saftner and Baker, unpublished results). This finding further supports the active nature of α-AIB transport into the vacuole. Influx into the cytoplasm may at times also be an active process in that at low applied concentrations of α-AIB (see Table 5 B), the cytoplasmic concentration becomes higher than that of the free space/external solution.

CONCLUSIONS

The results provide a model for transport and compartmentation of neutral amino acids in tomato pericarp tissue. α-AIB is a non-metabolized neutral amino acid widely used for uptake and transport studies, and there is growing evidence that various neutral amino acids are transported by the same or similar mechanisms (6, 8). However, there is no present evidence that methionine, ACC, or AVG utilize the same carriers, or compartment in the same manner as AIB. Assuming that transport and compartmentation of AIB and other neutral amino acids are similar, a number of points to emerge from this study are noteworthy. The bulk of AIB is pumped into the vacuole, although the concentration in the vacuole may or may not be higher than that in the cytoplasm depending on the time of incubation. The volume of the vacuole is much larger than the combined volumes of the other compartments. Boudet et al (4) working with protoplasts and vacuoles of sweet clover, determined that a number of amino acids were present to a large extent in the vacuole, and Guy and Kende (5) found that most of the ACC present in pea and broad bean protoplasts was localized in the vacuole. Our results are in agreement with their findings.

A second finding that may be of importance to studies on relative rates of protein and ethylene synthesis using labeled methionine, AVG, and other amino acids, is that the concentration of α-AIB in the cytoplasm was progressively lower with ripening, when the external concentration of α-AIB was low (25 μM). However, at the higher concentration of 200 μM, the cytoplasmic concentration of α-AIB increased somewhat with fruit development (Table 5). If methionine and AVG are transported and compartmentalized in a similar manner, then their concentrations in the cytoplasm would be much lower in ripe than in green cells, when applied at low concentrations. At higher concentrations of 200 μM and above in the incubation medium, their concentrations in the cytoplasm of ripe cells should be somewhat higher than in green cells. Since protein and ACC synthesis are mediated in the cytoplasm, concentrations of methionine and amino acids such as AVG should be similar at various developmental stages, in order that relative rates of synthesis and inhibition can be determined. From the present results, it appears that different rates of protein synthesis in green and ripe tissues could be partially due to differing cytoplasmic concentrations of ^{14}C-methionine when the external concentration of methionine is low. The insensitivity of ethylene production in ripe tomato tissue to 340 μM AVG, would not appear to be due entirely to low cytoplasmic concentration of AVG, if the transport and compartmentalization of AVG and AIB are similar. Whether or not that assumption is valid, remains to be determined.

A third finding in this study was that the rate constant of vacuolar efflux increased dramatically during development, indicating an increased passive permeability of the tonoplast to this compound and presumably other neutral amino acids. This is in contrast to the results of Vickery and Bruinsma (17) who found no increase in passive permeability of the tonoplast to K^+ during ripening of the tomato. The $t_{1/2}$ for AIB efflux from the tonoplast was still large at the ripe stage, i.e., 45 hr. The significance of an increased tonoplast permeability to neutral amino acids is not readily apparent. It might alter vacuolar ACC access to the active site of the ethylene-forming enzyme.

Lastly, the apparent free space increased from 74 to 96 μl/g from the immature green to red stage. While this is a 30 % increase, it does not represent a large volume of the tissue. At the immature green stage the free space volume was ca. 7% of the tissue volume, while at the red stage it was ca. 9%. This is in contrast to the report by Sacher (14) that banana tissue free space to sugars increased to essentially 100% of the tissue volume at the respiratory peak. Richmond and Biale (13) also reported a large increase in free space in avocado tissue during ripening, with 97% of amino acid uptake being confined to the free space after a 60-minute incubation. Further work is needed to determine the magnitude of free space in various ripening fruits, to determine if tomato is an exception to the rule.

REFERENCES

1. Aharoni N, Anderson JD, Lieberman M. 1979. Production and action of ethylene in senescing leaf disks: Effect of indole-acetic acid, kinetin, silver ion, and carbon dioxide. Plant Physiol. 64: 805-809.
2. Baker JE, Lieberman M, Anderson JD. 1978. Inhibition of ethylene production in fruit slices by a rhizobitoxine analog and free radical scavengers. Plant Physiol. 61: 886-888.
3. Baker JE, Anderson JD. 1982. Amino acid incorporation into proteins during ripening of the tomato fruit. Plant Physiol. Suppl. 69:42.
4. Boudet AM, Canut H, Alibert, G. 1981. Isolation and characterization of vacuoles from Melilotus alba mesophyll. Plant Physiol. 68: 1354-1358.
5. Guy M, Kende H. 1983. Ethylene formation in pea protoplasts. Plant Physiol. Suppl. 72: 38.
6. Jung KD, Luttge U. 1980. Amino acid uptake by a mechanism with affinity to neutral L- and D-amino acids. Planta 150: 230-235.
7. Kende H, Boller, T. 1981. Wound ethylene and 1-aminocyclopropane-1-carboxylate synthase. Planta 151:476-481.
8. Kinraide TB, Etherton B. 1980. Electrical evidence for different mechanisms of uptake for basic, neutral, and acidic amino acids in oat coleoptiles. Plant Physiol. 65: 1085-1089.
9. Luttge U, Higinbotham N. 1979. Transport in plants. Springer Verlag, New York.
10. Macklon AES, Higinbotham N. 1970. Active and passive transport of potassium in cells of excised pea epicotyls. Plant Physiol. 45: 133-138.
11. Mans R, Novelli GD. 1961. Measurement of the incorporation of radioactive amino acids into protein by a filter paper disk method. Arch. Biochem. Biophys. 94: 48-53.
12. Parr D, Edelman J. 1976. Passage of sugars across the plasmalemma of carrot callus cells. Phytochem. 15:619-623.
13. Richmond, A, Biale, JB. 1966. Protein and nucleic acid metabolism in fruits. 1. Studies on amino acid incorporation during the climacteric rise in respiration of the avocado. Plant Physiol. 41: 12337-1253.
14. Sacher, JA. 1966. Permeability. Characteristics and amino acid incorporation during senescence (ripening) of banana tissue. Plant Physiol. 41: 701-708.
15. Saftner, RA, Daie, J, Wyse, RE. 1983. Sucrose uptake and compartmentation in sugar beet taproot tissue. Plant Physiol 72:1-6.
16. Satoh, S, Esashi, Y. 1980. α-Aminoisobutyric acid: A probable competitive inhibitor of conversion of 1-aminocyclopropane-1-carboxylic acid to ethylene. Plant and Cell Physiol 21: 939-949.
17. Vichery, RS, Bruinsma, J. 1973. Compartments and permeability for potassium in developing fruits of tomato (Lycopersicon esculentum Mill.) J. Exper. Bot. 24: 1261-1270.

EFFECT OF SALINITY ON THE RIPENING PROCESS IN TOMATO FRUITS: POSSIBLE ROLE OF ETHYLENE

Y. MIZRAHI[1,2] and S. (MALIS) ARAD[2]

[1]Department of Biology, and

[2]The Institutes for Applied Research, Ben-Gurion University of the Negev, P.O. Box 1025, Beer-Sheva 84110, Israel

Exposure of tomato plants to salinity enhances the ripening process in their fruits. Thus, all the ripening parameters measured, e.g. CO_2 and ethylene evolution rates, pigment concentration, fruit softening, polygalacturonase activity, and taste, were higher than in their respective controls (4).

The nonripening tomato mutants *rin* and *nor* do not ripen according to all these ripening parameters (6). Ethylene application can induce some of the ripening processes in the *rin* tomato mutant (at least partially), while the *nor* tomato mutant is not responsive to the same treatment (6). However, exposure of the plants to salinity induces fruit ripening in *nor* but not in *rin*. Similar ripening behavior of *nor* was obtained by exposure to water stress induced by water shortage or kinetin treatment (1). The values of all the ripening parameters in the treated plants were nevertheless lower than in the normal cultivars (1, 5).

When the *nor* gene was introduced into five different genetic backgrounds and the plants were exposed to salinity, induction of ripening was evident in all the genotypes. However, the intensity of the ripening process varied among the various genotypes (Arad and Mizrahi, unpublished data). This phenomenon is reminiscent of other mutations, which show different degrees of expression on different genetic backgrounds. The question arises whether a correlation does exist between the degree of ripening of these mutants under saline

Y. Fuchs and E. Chalutz (eds.) Ethylene: Biochemical, Physiological and Applied Aspects.
ISBN 90-247-2984-X. Printed in The Netherlands
© 1984, Martinus Nijhoff/Dr W. Junk Publishers, The Hague.

conditions and their capacity to evolve ethylene.

Tomato plants (*Lycopersicon esculentum* Mill) of five different varieties, all homozygous for the nonripening gene *nor*, were kindly supplied by Prof. N. Kedar and collaborators from the Faculty of Agriculture of the Hebrew University of Jerusalem, Rehovot, Israel. These homozygous *nor* genotypes were grown in half-strength Hoagland solution which was changed weekly. On appearance of the first flowers, NaCl (3 g/l) was added to half of the plants while the remaining served as untreated controls. Fruits were hand-pollinated and tagged as previously described (2,3). When fruits reached various degrees of development they were harvested and either analyzed immediately or kept at -20°C until further analysis. Ethylene evolution rates were measured with fruits kept in closed jars in which humidifed air free of CO_2 and C_2H_4 was flowing at a rate of 100-200 ml/hr. Ethylene was analyzed by gas chromatography as previously described (5). For the pigment concentration analyses pericarp discs of the various fruits were extracted with an acetone/hexane mixture and analyzed as already described (5).

Fruits were harvested at various degrees of development, all beyond 100% of development, and the ethylene evolution rate was measured for each combination of genotype and treatment. Pigment concentration was also analyzed at the stage of development that correlated with highest ethylene evolution rates. The lycopene concentration was chosen to represent the degree of ripening. The degree of ripening in all the individual treatments, as expressed by the level of the red pigment (lycopene measured at 505 nm), was found to correlate significantly (p > 1%) with the ethylene evolution rates (at their peak levels) (Fig. 1).

Tomato fruits that reached at least a light red color evolved ethylene at more than 1 μl kg^{-1}, while tomato fruits that remained unripe (green or yellow without noticeable lycopene) evolved ethylene at a rate below 1 μl $kg^{-1}hr^{-1}$. Thus, this value is suggested as critical to the ripening process. Inasmuch as *nor* tomatoes usually do not ripen and

331

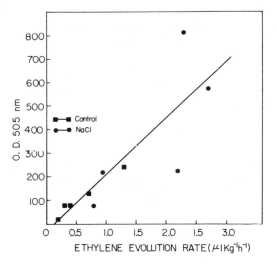

FIGURE 1. Correlation
between fruit pericarp
lycopene concentration
(O.D. 505 nm) and ethyl-
ene evolution rates in
tomatoes from plants
homozygous for *nor* on 5
different genetic back-
grounds, which were ex-
posed to salinity.
(r = 0.8295 at p > 1%;
y = 0.233X - 0.036)

and do not respond to exogenous ethylene but can be induced
to ripen by stress, they can serve as a model system for
understanding the ripening process.   That is,   the induction of ripening in the *nor* tomato mutant by salinity can
be used to determine the role of ethylene in ripening, i.e.,
is ethylene a symptom of ripening or its inducer?

REFERENCES

1. Arad(Malis) S, Mizrahi Y. 1983. Stress-induced ripen-
    ing of the nonripening tomato mutant *nor*. Physiol.
    Plant. 59:213-217.
2. Lyons JM, Pratt HK. 1964. Effect of stage of maturity
    and ethylene treatment on respiration and ripening of
    tomato fruits. J. Am. Soc. HortSci. 84:491-500.
3. McGlasson WB, Dostal HC, Tigchelaar EC. 1978. Compari-
    son of propylene induced response of immature fruit of
    normal and mutant *rin* tomatoes. Plant Physiol. 56:544-
    546.
4. Mizrahi Y. 1982. Effect of salinity on tomato fruit
    ripening. Plant Physiol. 69:966-970.
5. Mizrahi Y, Zohar R, Arad(Malis) S. 1982. Effect of
    sodium chloride on fruit ripening of the nonripening
    tomato mutant *nor* and *rin*. Plant Physiol. 69:497-501.
6. Tigchelaar, EC, McGlasson WB, Buescher RW. 1978.
    Genetic regulation of tomato fruit ripening. HortSci.
    13:508-513.

EFFECT OF MANNOSE ON ETHYLENE EVOLUTION AND RIPENING OF PEAR FRUIT

Chris B. Watkins and Chaim Frenkel, Department of Horticulture and Forestry, Rutgers University, New Brunswick, New Jersey 08903

As part of an investigation into the role of glycosidases in fruit ripening, we applied D(+) glucose, galactose and mannose as competitive inhibitors of their corresponding glycosidases. Preliminary experiments showed that mannose application inhibited softening and ethylene production. There are at least two possible explanations for this effect: (1) α-Mannosidase activity is inhibited by mannose, thus implicating this enzyme as having a controlling function in fruit ripening involved perhaps in glycoprotein processing; (2) Mannose is phosphosylated and not further metabolized, consequently lowering cellular inorganic phosphate levels and decreasing ATP synthesis. It is this mechanism by which the inhibitory effects of mannose (and its analogues, glucosamine and 2-deoxyglucose) on many metabolic processes have been explained (Herold and Lewis 1977, New Phytol 79:1-110).

We have been investigating the second possibility, and it appears that phosphate sequestration does not explain all our data:

(1) The mannose analogues, glucosamine and 2-deoxyglucose, did not fully mimic the action of mannose.

(2) Application of phosphate with mannose enhanced the mannose inhibition of softening and ethylene evolution rather than alleviating it.

(3) There was only a minor effect of mannose on respiration.

This work is still in progress, and analyses of adenylate, sugar phosphate and inorganic phosphate levels are incomplete. These should help to explain the phenomenon described here, and may further extend our knowledge on the relationships of carbohydrate metabolism to softening and ethylene evolution.

Y. Fuchs and E. Chalutz (eds.) Ethylene: Biochemical, Physiological and Applied Aspects.
ISBN 90-247-2984-X. Printed in The Netherlands
© 1984, Martinus Nijhoff/Dr W. Junk Publishers, The Hague.

P A R T I C I P A N T S

Abeles, F.B.	USDA-ARS, Appalachian Fruit Research Station, Box 45, Kearneysville, WV 25430, USA
Achilea, O.	ARO, The Volcani Center, Bet Dagan 50250, Israel
Aharoni, N.	ARO, The Volcani Center, Bet Dagan 50250, Israel
Amrhein, N.	Ruhr-Universität, Bochum 4630, Bochum-Querenberg, Germany
Anderson, J.D.	USDA-ARS, Plant Hormone Lab., BARC-West, Beltsville, MD 20705, USA
Apelbaum, A.	ARO, The Volcani Center, Bet Dagan 50250, Israel
Arad, S.	Research & Development Authority, Ben-Gurion University of the Negev, Be'er Sheva 84100, Israel
Baker, J.E.	USDA-ARS, Plant Hormone Lab., BARC-West, Beltsville, MD 20705, USA
Bangerth, F.	Universität Ohoenheim, 7000 Stuttgart (70), Germany
Ben-Arie, R.	ARO, The Volcani Center, Bet Dagan 50250, Israel
Ben-Tal, Y.	ARO, The Volcani Center, Bet Dagan 50250, Israel
Ben-Yehoshua, S.	ARO, The Volcani Center, Bet Dagan 50250, Israel
Beyer, E.M. Jr.	E.I. du Pont de Nemours & Co. Biochemicals Dept., Experimental Station, Wilmington, DE 19898, USA
Boller, T.	Botanisches Inst., Der Universität Basel, 4056 Basel, Switzerland
Borochov, A.	Faculty of Agriculture, Hebrew Univ. of Jerusalem, Rehovot 76100, Israel
Chalutz, E.	ARO, The Volcani Center, Bet Dagan 50250, Israel
Come, D.	Centre National de la Recherche Scientifique, Route de Gardes 92120, Meudon, France

Dagan, E.   Faculty of Agriculture, Hebrew Univ. of Jerusalem,
                           Rehovot 76100, Israel

De Proft, M.   Dept. of Biology,U.I.A. Universiteitsplein 1 B-2610,
                           Wilrijk, Belgium

Drori, A.   Faculty of Agriculture, Hebrew Univ. of Jerusalem,
                           Rehovot 76100, Israel

Epstein, E.   ARO, The Volcani Center, Bet Dagan 50250, Israel

Esquerré-Tugayé, M.T.   Université Paul Sabatier, 118, Route de Narbonne,
                           Toulouse, France

Faragher, J.   Horticultural Research Inst., Victorian Dept. of
                           Agriculture, P.O.B.174, Ferntree Gully, Vic. 3156,
                           Australia

Frenkel, C.   Dept. of Horticulture & Forestry, Rutgers - The
                           State University, Cook College, New Brunswick, NJ
                           08903, USA

Fuchs, Y.   ARO, The Volcani Center, Bet Dagan 50250, Israel

Gaash, D.   ARO, The Volcani Center, Bet Dagan 50250, Israel

Gepstein, S.   Dept. of Biology, Technion-Israel Inst. of Techno-
                           logy, Haifa, Israel

Ginzburg, C.   ARO, The Volcani Center, Bet Dagan 50250, Israel

Glazer, I.   ARO, The Volcani Center, Bet Dagan 50250, Israel

Goren, R.   Faculty of Agriculture, Hebrew Univ. of Jerusalem,
                           Rehovot 76100, Israel

Gorini, F.   IVTPA, Via Venezian 26 - 20133, Milano, Italy

Guy, M.   Michigan State Univ., MSU-DOE Plant Research Lab.,
                           East Lansing, MI 48824, USA

Hadar, A.   Faculty of Agriculture, Hebrew Univ. of Jerusalem,
                           Rehovot 76100, Israel

Hall, M.A.   Dept. of Botany and Microbiology, The University
                           College of Wales, Aberystwyth SY23 3DA, U.K.

Harshemesh, H.	ARO, The Volcani Center, Bet Dagan 50250, Israel
Hill, S.E.	Dept. of Botany, Royal Halloway College, Callow Hill, Virginia Water, Surrey, U.K.
Hobson, G.E.	Glasshouse Crops Research Institute, Littlehampton, West Sussex BN16 3PU, U.K.
Huberman, M.	Faculty of Agriculture, Hebrew Univ. of Jerusalem, Rehovot 76100, Israel
Hyodo, H.	Faculty of Agriculture, Shizuoka Univ., Ohya, Shizuoka 422, Japan
Itzhaki, H.	Faculty of Agriculture, Hebrew Univ. of Jerusalem, Rehovot 76100, Israel
Jaffe, M.J.	Wake Forest Univ., Winston-Salem, NC 27109, USA
Jayasingh, D.	Ministry of Agriculture, Hope Gardens, Kingston 6, Jamaica
Jona, R.	Istituto di Coltivazioni Arboree, Via P. Giura 15, 1-10126, Torino, Italy
Kahn, V.	ARO, The Volcani Center, Bet Dagan 50250, Israel
Kapulnik, E.	ARO, The Volcani Center, Bet Dagan 50250, Israel
Kende, H.	Michigan State Univ., MSU-DOE Plant Research Lab., East Lansing, MI 48824, USA
Keren-Paz, V.	Faculty of Agriculture, Hebrew Univ. of Jerusalem, Rehovot 76100, Israel
Landstein, D.	Ben-Gurion Univ. of the Negev, Be'er Sheva 84100, Israel
Lavee, S.	ARO, The Volcani Center, Bet Dagan 50250, Israel
Lederman, E.	Faculty of Agriculture, Hebrew Univ. of Jerusalem, Rehovot 76100, Israel
Lehrer, M.	Ben-Gurion Univ. of the Negev, Be'er Sheva 84100, Israel

Leshem, Y.Y.   Dept. of Botany, Bar-Ilan Univ., Ramat Gan, Israel

Lieberman, F.   107 Delford Avenue, Silver Spring, MD 20904, USA

Locker, A.   Tel-Aviv University, Ramat Aviv, Israel

Masia, A.   Istituto di Coltivazioni Arboree, Via Gradenigo 6,
                      35100, Padova, Italy

Mattoo, A.K.   USDA-ARS, Plant Hormone Lab., BARC-West, Beltsville,
                      MD 20705, USA

Mayak, S.   Faculty of Agriculture, Hebrew Univ. of Jerusalem,
                      Rehovot 76100, Israel

Meir, S.   ARO, The Volcani Center, Bet Dagan 50250, Israel

Mizrahi, Y.   Ben-Gurion Univ. of the Negev, Be'er Sheva 84100,
                      Israel

Mor, Y.   Ministry of Agriculture, Haqirya, Tel Aviv, Israel

Morgan, P.W.   Dept. of Plant Sciences. Texas A&M Univ., College
                      of Agriculture, College Station, TX 77843, USA

Nardin, K.   Versuchszentrum, 'Laimburg' 39040, Ora (Bolzano),
                      Italy

Native, D.   ARO, Newe Ya'ar Regional Experiment Station, P.O.
                      Haifa, Israel

Nichols, R.   Glasshouse Crops Research Inst., Worthing Rd.,
                      Rustington, Littelhampton, West Sussex, BN16 3PU,
                      U.K.

Nir, G.   ARO, The Volcani Center, Bet Dagan 50250, Israel

Nir, M.   Faculty of Agriculture, Hebrew Univ. of Jerusalem,
                      Rehovot 76100, Israel

Offer, R.   ARO, The Volcani Center, Bet Dagan 50250, Israel

Osborne, D.J.   Agricultural Research Council, Weed Research Org.,
                      Sandy Lane, Yarnton, Oxford, U.K.

Pesis, E.   ARO, The Volcani Center, Bet Dagan 50250, Israel

Philosoph-Hadas, S.	ARO, The Volcani Center, Bet Dagan 50250, Israel
Plesner, O.	Faculty of Agriculture, Hebrew Univ. of Jerusalem, Rehovot 76100, Israel
Preger, R.	Dept. of Biology, Technion-Israel Inst. of Technology, Haifa, Israel
Rappaport, L.	Univ. of California, Davis, CA 95616, USA
Rasmussen, G.	USDA, European Marketing Research Center, Marconi-Straat 38b, 3029 AK Rotterdam, The Netherlands
Retig, N.	BARD, The Volcani Center, Bet Dagan 50250, Israel
Reznisky, D.	Israel Fruit Growers Association, Cold Storage Res. Lab., Qiryat Shemona, Israel
Riov, J.	Faculty of Agriculture, Hebrew Univ. of Jerusalem, Rehovot 76100, Israel
Sagee, O.	Faculty of Agriculture, Hebrew Univ. of Jerusalem, Rehovot 76100, Israel
Schiffmann-Nadel, M.	ARO, The Volcani Center, Bet Dagan 50250, Israel
Schnell, Mr. & Mrs. S.	87-18 Santiago Street, Holliswood, NY 11423, USA
Shilo, M.	BARD, The Volcani Center, Bet Dagan 50250, Israel
Shilo, R.	Faculty of Agriculture, Hebrew Univ. of Jerusalem, Rehovot 76100, Israel
Shitrit, Y.	ARO, The Volcani Center, Bet Dagan 50250, Israel
Sisler, E.C.	Dept. of Biochemistry, North Carolina State Univ. Box 5050, Raleigh, NC, USA
Sive, A.	Israel Fruit Growers Association, Cold Storage Res. Lab., Qiryat Shemona, Israel
Smith, A.R.	Dept. of Botany and Microbiology, The University College of Wales, Aberystwyth SY23 3DA, U.K.
Spiglstein, H.	Faculty of Agriculture, Hebrew Univ. of Jerusalem, Rehovot 76100, Israel

Stead, A.D.   Dept. of Botany, Royal Halloway College, Callow Hill, Virginia Water, Surrey, U.K.

Suttle, J.C.   USDA, State University Station, Fargo, ND 58105, USA

Teitel, D.   Ben-Gurion Univ. of the Negev, Be'er Sheva 84100, Israel

Temkin-Gorodeiski, N.   ARO, The Volcani Center, Bet Dagan 50250, Israel

Thompson, J.E.   Dept. of Biology, University of Waterloo, Waterloo, Ontario, Canada

Tirosh, Z.   Faculty of Agriculture, Hebrew Univ. of Jerusalem, Rehovot 76100, Israel

Trabitsh, T.   ARO, The Volcani Center, Bet Dagan 50250, Israel

Van Loon, L.C.   Dept. of Plant Physiology, The Agricultural Univ., 6703 BD Wageningen, The Netherlands

Van Meeteren, U.   Bulb Research Centre, P.O.B. 85, 2160 AB Lisse, The Netherlands

Vardimon, V.   Ben-Gurion Univ. of the Negev, Be'er Sheva 84100, Israel

Veen, H.   Centre for Agrobiological Research, Bronsesteeg 65, P.O.B. 14, 6700 AA Wageningen, The Netherlands

Villa, I.   "Isolcell", Via A. Meucci, Laives, Bolzano, Italy

Viner, L.   ARO, The Volcani Center, Bet Dagan 50250, Israel

Vinkler, C.   ARO, The Volcani Center, Bet Dagan 50250, Israel

Wallenstein, I.S.   ARO, The Volcani Center, Bet Dagan 50250, Israel

Waltering, E.J.   Rosendaalsestraat 320, Arnhem, The Netherlands

Watkins, C.B.   Dept. of Horticulture & Forestry, Rutgers - The State University, New Brunswick, NJ 08903, USA

Wineberg, D.   Ben-Gurion Univ. of the Negev, Be'er Sheva 84100, Israel

Yang, S.F.   Vegetable Crops Dept., University of California,
                            Davis, CA 95616, USA

Zach, A.   Faculty of Agriculture, Hebrew Univ. of Jerusalem,
                            Rehovot 76100, Israel

Zauberman, G.   ARO, The Volcani Center, Bet Dagan 50250, Israel

I N D E X

Dianthus caryophyllus, see
  Carnation
Diamine oxydase, 155
Digitalis, 107
Disease symptoms, 171
Dormancy, 163
Driselase, 191

Ecbalium elaterium, 222
Elicitor, 199, 217
  fungal, 186
Epinasty, 162, 171
Ethephone, 107, 137, 163, 165,
    173, 200, 255, 279, 283
Ethrel *see* Ethephone
Ethylene
  absorption, 130
  action, 68, 75, 104, 235, 275,
    303
  analogues, 55
  antagonists, 284
  binding, 45, 99
    inhibition of, 99
  binding site, 45, 55, 282
  biosynthesis, 1, 21, 33, 95, 97,
    105, 111, 123, 129, 159, 161,
    165, 181, 236, 261, 277, 291,
    317, 333
    autocatalytic, 4, 103, 187,
    282
    effect of carbohydrates, 95,
    130
    cellulysin-induced, 183
    inhibition of, 3, 164, 169
    membrane involvement in, 139,
    186

Ethylene (continued)
  biosynthesis
    regulation of, 129
    wound induced, 93, 203, 232
    forming enzyme (EFE), 1, 93, 145,
      163, 186
    glycol, 65
    growth response, 121, 255
    IAA induced, 129
    inducing factor, 189, 217
    involvement in disease, 171
    metabolism, 65, 76, 103
    monooxygenase, 75
    oxidation, 75, 236
    oxide, 65, 75
    regulator of ripening, 281, 291,
      329
    role in abscission, 221, 231
    saturation of, 81
    sensitivity to, 238
    target cell, 231
    treatment, 149, 181, 246
    uptake of, 80
Eucalyptus, 268

Ficus lyrata, 161
Flax, 48
Flower, 101, 163, 167, 234
Flowering, 165
Free radical, 2, 39, 113
  Scavenger of, 21, 105
Free space volume, 318
  apparent, 326
  of cell wall, 322
Fruit
  abscission, 234